# THE BIOMIMICRY
# REVOLUTION

# THE BIOMIMICRY
# REVOLUTION

*Learning from Nature
How to Inhabit the Earth*

HENRY DICKS

*Columbia University Press*
*New York*

Columbia University Press
*Publishers Since 1893*
New York   Chichester, West Sussex
cup.columbia.edu
Copyright © 2023 Columbia University Press

Library of Congress Cataloging-in-Publication Data
Names: Dicks, Henry, author.
Title: The biomimicry revolution: learning from nature how to inhabit the earth.
Description: New York: Columbia University Press, 2023. | Includes bibliographical
    references and index.
Identifiers: LCCN 2022025108 (print) | LCCN 2022025109 (ebook) | ISBN 9780231208802
    (hardback) | ISBN 9780231208819 (trade paperback) | ISBN 9780231557634 (ebook)
Subjects: LCSH: Biomimicry—Philosophy. | Bionics—Philosophy. | Environmentalism. |
    Nature. | Knowledge, Theory of.
Classification: LCC TA164.2 .D53 2023 (print) | LCC TA164.2 (ebook) |
    DDC 003/.5—dc23/eng/20221006
LC record available at https://lccn.loc.gov/2022025108
LC ebook record available at https://lccn.loc.gov/2022025109

Cover design: Milenda Nan Ok Lee
Cover photo: Andrew Bassett @ Shutterstock

*For Patrick*

# Contents

# *Preface*

Philosophy is the love (*philo*) of wisdom (*sophia*). Biomimicry is the imitation (*mimesis*) of life (*bios*). This book explores the idea that there is a profound link between the two, that wisdom cannot arise through human genius alone, for it also requires us to learn from the genius of nature. A true lover of wisdom, from this perspective, would be a lover of nature, and of the invaluable lessons she has to offer.

As it stands, however, the links between philosophy and biomimicry have been little explored. The biomimicry movement is predominantly practical in orientation. Faced with what is now widely recognized as an ecological emergency, its main aims have been to promote biomimicry as a powerful response to that emergency, to develop tools and methodologies that facilitate biomimicry's application to concrete problems in engineering and design, and—last but not least—to develop concrete solutions to these problems. The more philosophical aspects of biomimicry have been largely overlooked.

Conversely, it is not an understatement to say that mainstream philosophy has barely heard of biomimicry. There are exceptions, notably the pioneering work of Freya Mathews[1] and Bernadette Bensaude-Vincent,[2] as well as important recent contributions from the likes of Vincent Blok[3] and Hub Zwart.[4] But, because of biomimicry's practical orientation, and because of obstacles posed by the structure and content of contemporary philosophy, biomimicry has not yet made any significant impact on mainstream philosophy.

The overall result is that biomimicry today finds itself in a comparable situation to that of conservation in the late 1940s, as described by the great American ecologist Aldo Leopold. In his seminal 1949 essay "The Land Ethic," Leopold wrote: "The proof that conservation has not yet touched [the] foundations of conduct lies in the fact that philosophy and religion have not yet heard of it. In our attempt to make conservation easy, we have made it trivial."[5] The same is true today of biomimicry. Philosophy has barely heard of it; and, in trying to make it accessible and easy, it all too easily comes to appear as unworthy of profound reflection.

If what Janine Benyus calls the "biomimicry revolution" is to succeed, this must change. If the practical ambitions of biomimicry are to be realized, it can only be on the basis of sustained engagement with its philosophical foundations. But, in order to embark on this undertaking, we must first take a closer look at what biomimicry is. Under the guises of biomimetics and bionics, imitating nature has been an established approach to technological innovation since the late 1950s and today plays a major role in such diverse fields as robotics, AI, nanotechnology, medicine, materials science, and architecture. Velcro, to give a classic example, was invented by the Swiss engineer George de Mestral, by imitating the way the burrs of the burdock thistle attached to the fur of passing animals. And, to give a topical example, many vaccines work by partially imitating the forms of viruses— but not their debilitating functions—so that the immune system is able to recognize and defend itself from the pathogens themselves.

But it is only since 1997, when Benyus started promoting "biomimicry," as something at least partially distinct from biomimetics, in her best-selling book *Biomimicry*, that it has been widely recognized that nature may also hold invaluable ecological lessons about how we might enduringly inhabit the earth.[6] If one considers the earth as a whole, it is clear that huge swathes of it have been destroyed, as natural ecosystems are replaced with artificial—especially agricultural, industrial, and urban—systems. But these artificial systems work in ways that are radically different from nature, and, in so doing, they go against and threaten both the ever-dwindling fragments of nature that remain and the global ecosystem to which they belong. Viewed in this light, the basic aim of biomimicry is to transform these artificial systems such that, in certain key ecological respects, they come to more closely resemble the natural ones they have replaced and may thus fit sustainably into the earth system as a whole.

Benyus achieved much more, however, than simply showing the significance of biomimicry to overcoming the ecological crisis, for—and this is the central thesis of this book—what she ultimately put forward was a *new philosophy*. The biomimicry revolution, from this perspective, is not only a technological revolution, even one of comparable scope and scale to the industrial revolution, but also a philosophical revolution concerning the very foundations of human existence. And yet, being a scientist, not a philosopher, Benyus did not explore and develop the philosophical aspects of her vision. Profound as her vision may have been, it was illustrated by inspiring examples rather than rigorously analyzed and developed at the level of its philosophical foundations. This, then, is the basic aim of the present book: to present, analyze, and explore the philosophical foundations of the biomimicry revolution.

Before I begin, let me first say a few words of thanks to those without whom this book would not have been possible. Its underlying environmental philosophy dates back to the work on Heidegger and contemporary French philosophy I carried out during my doctorate at the University of Oxford, and I would like to thank my doctoral supervisors, Christina Howells and Ben Morgan, for their contributions to this formative period of my intellectual development. As for my specific interest in biomimicry, it arose during two postdocs I carried out at the University of Lyon, financed by Suez Environment and Intelligences des Mondes Urbains (IMU), respectively, during which my thinking on the subject benefited from discussions with Jean-Philippe Pierron, Jean-Luc Bertrand-Krajewski, Christophe Ménézo, Claire Harpet, and Yvan Rahbé, colleagues of mine on a research project on biomimetic cities. I would also like to extend special thanks to my international colleagues in the philosophy of biomimicry, Freya Mathews, Bernadette Bensaude-Vincent, Hub Zwart, and especially Vincent Blok, not only for their various collaborations over the past few years—in particular on a special issue of *Environmental Values*—but also for their intellectual company in what would otherwise have been a very lonely research field.[7]

The bulk of this book was written during a visiting research fellowship at the Centre for the History and Philosophy of Science (HPS) of the University of Leeds under the mentorship of Ellen Clark, the centre's director. The pandemic meant that I spent less time at HPS than I would have liked, but I remain very grateful to Ellen both for supporting my research and for allowing me to benefit from HPS's wonderful virtual research environment. I would also like to thank Wendy Lochner, my editor at Columbia University Press, both for having taking on the project, and for her calm, reassuring, and relaxed approach, which has made the whole publication process very stress-free. I would also like to thank my two reviewers for showing such strong support of the project while also providing some very helpful suggestions for potential improvements.

Finally, at a personal level, I would like to thank my wife, Tihan, for supporting me so patiently throughout a project that has taken a little longer than planned, as well as my parents, Joan and Geoff, both for the unconditional love and support they have provided throughout the years, and for all the time they spent looking after our son, Patrick, during multiple COVID lockdowns and school holidays, thereby freeing up much needed writing time.

# THE BIOMIMICRY
# REVOLUTION

# Introduction

## Biomimicry as a New Philosophy

Biomimicry is an original and powerful response to the ecological crisis. Whereas traditional environmentalism seeks to limit the destruction of nature, primarily through the actions of preservation and conservation, biomimicry seeks to imitate nature in the design of artificial products and systems, to emulate nature in embracing an ecological way of being, and to learn from nature's hidden reserves of knowledge and wisdom.

As it stands, however, the philosophical aspects of this revolution have yet to be theorized in a comprehensive and rigorous manner. A number of popular science books have been published on the subject—in particular Benyus's *Biomimicry*, which first introduced the idea of a "biomimicry revolution"—but these typically give short shrift to the theoretical and philosophical aspects of biomimicry, preferring to concentrate on concrete examples of successful biomimetic innovation.[1] Something similar holds for the tens of thousands of scientific articles in biomimicry and related fields, the principal difference being that technical, scientific discourse takes the place of inspirational storytelling. The overall result, as environmental philosopher Freya Mathews explains, is that biomimicry "is still relatively philosophically underdeveloped, descriptive, and ad hoc in its approach and accordingly piecemeal in its results."[2] Lacking a cogent theoretical and philosophical basis, biomimicry has been unable to realize its revolutionary potential.

This situation has begun to change over the last decade, as environmental philosophers and philosophers of technology have begun to grapple with the basic questions and issues raised by biomimicry. How does biomimicry understand and relate to nature? How does biomimetic innovation work, and what exactly falls under its scope? In addition to providing models for new technologies, can nature also provide the basis of an ecological ethics? And how exactly are we to understand and put into practice the biomimetic view of nature as a source of knowledge and wisdom?

These philosophers have also sought to situate the biomimicry revolution relative to the Western intellectual tradition. Unlike the popular and academic scientists, it has not escaped their notice that, at the very origin of that tradition in ancient Greece, it was commonplace to view technology (*technē*) as imitation (*mimesis*) of nature (*physis*).[3] Likewise, the proximity between biomimicry and Eastern philosophical traditions, especially Daoism, has received some much deserved attention.[4] But, like the work of the popular and academic scientists, the philosophical study of biomimicry remains somewhat piecemeal. Important questions have been raised, and some tentative answers proposed, but a comprehensive philosophical analysis of biomimicry has yet to be undertaken. This is reflected in the nature of philosophical publications on the subject. Despite the steadily growing corpus of academic articles, a book-length philosophical study of biomimicry has yet to appear—at least until now.

As with all new endeavors, the question arises as to where and how to begin. I propose to begin at what I take to be the beginning, Benyus's *Biomimicry*, for it here that both the expression and the concept of a biomimicry revolution first emerged. Indeed, not only did Benyus coin this expression but, as noted above, she was also the first to think cogently about biomimicry as both a strategy for technological innovation and—even primarily—as a revolutionary response to the ecological crisis. To begin with Benyus's book does, however, pose the already mentioned difficulty that she gives short shrift to the theoretical and philosophical aspects of biomimicry. But, while short, some of the shrift she does give is of great depth and significance. In a brief epigraph to her book, Benyus puts forward three fundamental principles, which, I maintain, together account for biomimicry's revolutionary character:

i.   *Nature as model*. Biomimicry is a new science that studies nature's models and then imitates or takes inspiration from these designs and processes to solve human problems (e.g., a solar cell inspired by a leaf).

ii.  *Nature as measure*. Biomimicry uses an ecological standard to judge the "rightness" of our innovation. After 3.8 billion years of evolution, nature has learned: What works. What is appropriate. What lasts.

iii.   *Nature as mentor.* Biomimicry is a new way of viewing and valuing nature. It introduces an era based not on what we can *extract* from the natural world, but on what we can *learn* from it.[5]

Benyus may have relatively little else to say about these principles, and yet it is in them, I believe, that the ultimate foundations of the biomimicry revolution lie.

It would be a mistake, however, to think that Benyus simply invented these principles. The first principle, nature as model, was already present in technological innovation long before Benyus wrote *Biomimicry.* When she says that biomimicry "imitates or takes inspiration" from nature, this may be taken as a direct reference to two preexisting fields of technological research and development, biomimetics and bioinspiration, the first of which covers imitating nature and the second taking inspiration from nature. Such is the overlap of these fields with biomimicry that they are sometimes taken to be synonymous with it.[6]

To see biomimicry as synonymous with biomimetics and bioinspiration is, however, to ignore the fact that it is only in biomimicry that one finds the principle of nature as measure, and therewith the idea that nature provides an "ecological standard" against which the "rightness" of our innovations is to be judged. This second principle is of quite different origin; it would appear to derive, rather, from the work of the contemporary American agroecologist Wes Jackson. Just three years before Benyus's *Biomimicry,* Jackson published an essay, "Nature as Measure," in which he criticized the singular focus of traditional environmentalism on preserving wild nature and advocated measuring human technological products and systems against nature's ecological standard.[7] Benyus explicitly cites this essay in the first chapter of her book, which documents Jackson's attempt to farm according to this principle.[8]

As for the third principle, nature as mentor, it is present in biomimetics and bioinspiration as well as in the work of Jackson. In academic discourses, it figures primarily as a rather casual trope (e.g., the "school of nature",[9] "lessons from nature"),[10] while in Jackson it takes on a more thoughtful and serious character (e.g., "a return to nature as our primary teacher").[11] That the principle of nature as mentor is present in all of these areas is not, however, surprising, for whether one takes nature as model or as measure, one is engaged in learning from nature.

We may conclude from this that Benyus's seminal exposition of biomimicry arose from bringing the principle of nature as model already present in biomimetics and bioinspiration together with the ecologically oriented principle of nature as measure discovered in the work of Jackson, while at the same time recognizing that implicit in these two principles is a third principle, nature as mentor, which amounts to a "new way of viewing and valuing nature" capable of introducing a new "era."

But, while these three principles may underpin the biomimicry revolution, there is an important question on which they depend but leave answered: What is this thing—nature—that we are to take as model, measure, and mentor? Philosophical inquiry into biomimicry, it would seem, cannot aim just to analyze and explore Benyus's three principles; it must also seek an answer to a deeper question—the question of the nature of nature—without which these principles could not get off the ground. The philosophical study of biomimicry, it follows, must first examine the question of what nature is, before proceeding to a consideration of Benyus's three principles, which it must not only analyze individually but also articulate with one another and with the understanding of nature on which they depend in such a way that there may emerge a fuller and more comprehensive understanding of the biomimicry revolution's philosophical foundations.

## FOUNDATIONS OF A NEW PHILOSOPHY

In 2016, I published an article, "The Philosophy of Biomimicry."[12] Despite its many weaknesses and limitations, it had the merit of discussing each of the three fundamental principles of biomimicry, while also putting forward an answer to the question of the nature of nature. Inasmuch as the present work adopts this same approach, it may be seen as an expansion of this early article into a much more comprehensive book-length study. There is, however, an important yet subtle difference with my earlier article. The title of the article apprehends biomimicry *as an object of philosophical study*. The philosophy of biomimicry, it would seem, is in this respect no different from, say, the philosophy of biology, for both study some preexisting field of human activity with a view to finding answers to the philosophical questions it raises. The present book, by contrast, apprehends biomimicry *as itself a new philosophy*. Its aim, therefore, is not so much to undertake philosophical inquiry into biomimicry, understood as one object of philosophical inquiry among others, as to think through the foundations of biomimicry, understood as itself a new philosophy.

If I take biomimicry to be a new philosophy, it is on account of its fundamental principles. Further, it is with respect to these fundamental principles that biomimicry is quite different from, say, biology. In biology, one finds various unresolved conceptual and theoretical problems, questionable beliefs and presuppositions, methodological disagreements, and so on; and it is these that philosophers of biology take as their objects of their study. But biology is a science, not a philosophy, and, in keeping with this, there is in biology no equivalent to the three fundamental principles of biomimicry, no concise set of fundamental philosophical principles that together underlie and structure it. But if my inquiry into the three fundamental principles of biomimicry is at the same time an inquiry into a new philosophy, we must first inquire briefly into what philosophy is, such that it may become more readily apparent why these principles may be seen to lay the foundations of a *new* philosophy.

As a philosophy undergraduate at University College London, I was told that my discipline had three main branches: metaphysics and epistemology, ethics, and aesthetics. This tripartite structure is, of course, something of a simplification, for there are many other branches of philosophy. But these other branches may for the most part be assimilated or attached to these three basic ones. Philosophy of mind and philosophy of science may be assimilated or attached to metaphysics and epistemology, political philosophy to ethics, and the philosophy of art to aesthetics. It is also important to note that this tripartite structure is broadly Kantian, for it corresponds more or less to the division of philosophy set forth by Immanuel Kant in his three critiques: the *Critique of Pure Reason* deals with metaphysics and epistemology; the *Critique of Practical Reason* with ethics; and the *Critique of Judgment* with, among other things, aesthetics.

There are two major ways in which my view of the basic structure of philosophy differs from the broadly Kantian one that prevails today. The first has to do with metaphysics and epistemology. Kant saw metaphysics as concerned with purely rational inquiry into things that cannot be perceived by the senses, such as God, the soul, or the origin of the cosmos. Knowledge of these things, he argued, was impossible, because knowledge is only possible through the application of our in-built faculty of understanding to those entities that appear to the senses. The consequence of this Kantian revolution is that epistemology, the study of knowledge, has to a large extent replaced metaphysics as the fundamental branch of

1. Epistemology (and metaphysics)    2. Ethics    3. Aesthetics

**0.1** The present structure of philosophy

philosophy (see fig. 0.1). Metaphysics may continue, but it lacks the proud status of "first philosophy" that it held before Kant.

Where my view of philosophy breaks with that of Kant is that, while accepting his rejection of metaphysics, it seeks to rehabilitate another branch of philosophy that Kant identifies with metaphysics—namely, ontology, the study of being. So, whereas Kant rejected ontology, on the grounds that there can be no knowledge of "things in themselves" (noumena) but only of "things as they appear to us" (phenomena), I propose to rehabilitate ontology, which must henceforth be seen as something radically different from metaphysics.[13] In this manner, ontology—which can no longer be seen in a Kantian manner as metaphysical inquiry into "things in themselves"—may emerge as an area of philosophical inquiry in its own right.

Ontology, from this perspective, is not just another area of philosophy but, rather, its very foundation, for it is from this foundation that all the other areas branch out. The tripartite structure derived from Kant may thus be replaced with a quadripartite structure, the foundation of which lies in that branch of philosophy that Kant rejected along with metaphysics: ontology. Further, since ontology forms the base of this quadripartite structure, it is no longer associated specifically or primarily with epistemology; on the contrary, it becomes the foundational area of philosophical inquiry, albeit one whose relation to the other areas of philosophy—not just epistemology—calls for further investigation.

The second difference from the prevailing, broadly Kantian, division of philosophy is of more direct relevance to biomimicry. To understand this difference, let us consider Kant's third critique. Undoubtedly the best-known part of the *Critique of Judgment* is the first part, "the critique of aesthetic judgment," especially the analytic of the beautiful and the sublime. But the third critique also contains a second part, the "critique of teleological judgment," which is concerned with judgments relating to the presence of purpose (telos) in what Kant calls "organized nature." The overarching concern of Kant's third critique, we may conclude, is not aesthetics. Further, it is not only true that the third critique is only partly concerned with aesthetics but also that the first critique is concerned to a significant extent with aesthetics, especially the famous "transcendental aesthetic," in which Kant discusses aesthetics in the broad sense of *aesthesis*, the Greek word for "sensibility."

In view of this, it may perhaps be thought that what holds the third critique together, and thus accounts for its unity, is not aesthetics, but, rather, as its title suggests, the notion of judgment. But, as the overarching concept of Kant's philosophy as a whole, judgment is equally present in the first and the second critique. The difference between the three critiques lies not in the fact that the third critique alone is concerned with judgment, but in the fact that it is concerned with different types of judgment, aesthetic and teleological judgment, instead of theoretical (first critique) and practical (second critique) judgment.

So what does account for the unity of the third critique? If the first critique is concerned with epistemology (and the rejection of metaphysics), and the second with ethics, with what is the third critique concerned? The answer, I suggest, is that it is concerned with what Kant calls *Technik*, a German word that would in most contemporary contexts be translated as "technology," but that I propose to translate as "technics." When analyzing organized nature, Kant understands it by analogy with human technics, and thus as what he calls *Technik der Natur* (technics of nature or natural technics). Although nature is not purposive, when we seek to understand the various different phenomena present in organized nature (especially living beings), we cannot but see them as purposive and so to understand them by analogy with human technics. When, for example, biologists study the wings of birds, they necessarily make the teleological judgment that, like the wings of planes, they serve the purpose of flying.

As for Kant's discussion of aesthetics, the concept of Technik is also of central importance here, for it underpins his understanding of the beautiful and the sublime. The judgment that an object such as a flower or a horse is beautiful, Kant tells us, is based on our awareness of what he calls "mere formal purposiveness," or "purposiveness without a purpose," by which he means that the form of the object appears purposive, as if it were a product of art, even though in our aesthetic appreciation of the object no actual purpose is ascribed to it.[14] The judgment that an object such as a storm or a jagged mountain range is sublime, by contrast, is based on our awareness of something devoid of form, limitless, and from which all purposiveness—and all possible analogy with art—is absent.[15] The concept of technics, we may conclude, is the guiding thread that ties together Kant's discussion of both aesthetic and teleological judgment, for in both cases a judgment is made regarding the apparent presence (or absence, in the case of the sublime) of purpose and design in nature. Further, it is also important to note that, while the primary focus of the third critique is what Kant calls "natural technics," he also engages in extensive discussions of art and technology, or "human technics," as one might call it. When we further note that discussions of both natural and human technics are more or less unique to the third critique, there can be little doubt that technics is its distinctive concern.[16]

Drawing on this reading of Kant's third critique, I suggest that we no longer see aesthetics as one of the basic branches of philosophy and that we replace it with technics, understood here as that branch of philosophy that is concerned with both "natural technics"—that is to say, apparently purposive design in nature, and "human technics," or purposive design on the part of humans—and thus also art and technology.

As a consequence of this change, I also propose that we reverse the order in which we consider the three secondary branches of philosophy. In the current division, there can be little doubt that epistemology (and metaphysics) is accorded greater significance than ethics, and ethics greater significance than aesthetics, for what is of most significance to contemporary philosophy is gaining knowledge about the world, with less significance being accorded to deciding on courses of right action in the world, and least significance of all to the aesthetic appreciation of the world. When we replace aesthetics with technics, however, the plausibility of this ordering is called into question. If it is easy to see aesthetics as of relatively little significance on account of its lack of practicality or functionality, the same cannot be said for natural and human technics, for without natural technics there would be no living beings, and without human technics there would be no human beings (assuming art and technology to be essential to humans).

This is not to say that technics is of greater significance than ethics or epistemology. If I propose to place technics before ethics and ethics before epistemology, it is not because this reflects a hierarchy of importance, but simply because, for methodological reasons that will only become apparent later on, it makes most sense to consider technics, ethics, and epistemology *in that order*. Philosophy, we may conclude, is primarily thinking about being (ontology), and secondarily thinking about making and producing (technics), right action (ethics), and knowledge (epistemology).

In light of this view of the basic structure of philosophy (see fig. 0.2), we are now in a position to see why I earlier claimed that biomimicry is a philosophy:

Technics    Ethics    Epistemology
\ ↑ /
Ontology

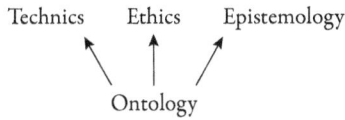

**0.2** A new structure for philosophy

there is a clear correspondence between the four basic areas of philosophy and the four basic areas of biomimicry. To inquire philosophically into the "nature of nature" is, at least potentially (i.e., if one identifies being with nature), to inquire into being. Under the prevailing post-Kantian division of philosophy, however, this inquiry has become impossible, for, after Kant, only science and not philosophy may inquire into nature. Philosophy may accompany science in its inquiry into nature, but it may not undertake any such inquiry on its own. Once ontology is rehabilitated, however, the possibility arises of a genuinely philosophical inquiry into being or nature and therewith of a new ontology. And since this new ontology may in turn underpin the three principles of biomimicry set forth by Benyus, it may also serve as an *ontology for biomimicry* (see chapter 1).

As for "nature as model," to inquire into this principle is clearly to undertake an inquiry into technics, for this principle is concerned with both of the branches of technics set out above: the natural and the human. Again, however, if one accepts the current structure of philosophy, it is practically impossible to think philosophically about the principle of nature as model. As technology assumes ever greater significance in human lives, philosophy of technology is certainly attracting more attention. And yet, viewed in relation to the prevailing structure of contemporary philosophy, it occupies a *philosophical no-man's-land*. It is sidelined from epistemology, for whatever knowledge it involves is not the representational knowledge about objective reality with which epistemology is for the most part concerned. It is sidelined from ethics, for technology is considered but a means to an end, and ethics is concerned mainly with ends. And it is sidelined from aesthetics, for since works of technology are primarily functional, whatever aesthetic qualities they may possess are of only secondary importance. As for the philosophy of biology, it is today but a subbranch of the philosophy of science, itself but a subbranch of epistemology (albeit an important one).

So, whereas, given the current structure of philosophy, to think philosophically about the technological imitation of nature requires one audaciously to bring together the philosophical no man's land that is philosophy of technology with that subbranch of a subbranch of epistemology that is philosophy of biology, in the new structure of philosophy that I propose, one may think about the technological imitation of nature in the context of a *basic branch of philosophy*: technics. Further, the technological imitation of nature is not just thinkable within the field of technics, for it proposes a way in which the two basic branches of technics—the natural and the human—may be articulated: through taking the former as model for the latter. The principle of nature as model, we may conclude, lies at the basis of a new technics: *biomimetic technics* (see chapter 2).

Turning now to the principle of nature as measure, to adopt an ecological standard as a measure of the rightness of at least some of our actions is clearly a normative proposal, for a norm is a standard against which the rightness of something is measured. Whether or not this normative proposal amounts to an ethics, whether the normativity in question is of the ethical variety, is another matter, which I will address later. But, thanks to the explicit reference here to a "standard" against which at least some forms of right action may be judged, the possibility emerges that a bona fide ethics could potentially be established on the basis of the principle of nature as measure. To take nature as measure would, in this scenario, also be to adopt a new ethics: *biomimetic ethics* (see chapter 3).

Lastly, it is not difficult to see that the principle of nature as mentor pertains to epistemology. Epistemology is concerned with the study of knowledge—its nature, its origin, its limits, and so on. What the principle of nature as mentor tells us is, first, that knowledge is already present in nature, and, second, that we may learn from that knowledge—a view that runs counter to the prevailing assumption that knowledge is generated by and present only in human beings, and that, to the extent that this knowledge concerns nature, nature is only ever its object, something *about* which humans may learn, but not its source, something *from* which they may learn. To take nature as mentor, it follows, is also to adopt a new epistemology: *biomimetic epistemology* (see chapter 4).

In view of this, it should now be abundantly clear why I view biomimicry as a new philosophy and not simply as another object of philosophical inquiry. Unlike other objects of philosophical inquiry, such as biology, music, or mathematics, it takes a stance with respect to each of the four basic branches of philosophy; in the form of its three fundamental principles, it puts forward a technics, an ethics, and an epistemology, and, since these principles all depend on a certain understanding of nature, it calls for a theory of being or nature, and thus an ontology, on which these principles may come to rest.

To say that biomimicry is a new philosophy must, however, be substantially qualified. Biomimicry, as will presently become apparent, is a specific type of philosophy—namely, an environmental philosophy. The basic concern of environmental philosophy is the human relationship to nature, where nature is viewed in a way that is relevant to the environmental crisis. Given that biomimicry is a new philosophy that rethinks both nature itself and the human relation to nature in response to the environmental crisis, it is not difficult to see that it is an environmental philosophy. It does, however, differ significantly from environmental philosophy as it exists today. At present, environmental philosophy is dominated by environmental ethics. In keeping with this, the principal claim discussed in contemporary environmental philosophy is that what must change in the human relationship to nature is our ethics, which should no longer center solely on humans but also, and perhaps even rather, on nature. The key concept of "nonanthropocentrism," to which this line of thinking has given rise, has not, however, exerted much influence outside of academic philosophy.[17]

In keeping with its current focus on environmental ethics, environmental philosophy has had relatively little to say about what I suggest we view as its other main branches: environmental ontology, environmental technics, and environmental epistemology. For the most part, views of nature present in environmental philosophy have been taken over directly from science, and, to the extent that the ontological dimension of the human relationship to nature has been addressed, it is typically only in rather vague and banal ways—for example, by saying that we mistakenly consider ourselves to be separate from nature, when we are in fact a part of it. This is not to deny that environmental philosophers spend a lot of time discussing the word and concept of nature, but these discussions—especially those concerning the presence of intrinsic value in nature—have for the most part been subsidiary to discussions in environmental ethics, and, as such, would perhaps be better categorized as "environmental meta-ethics," rather than as "environmental ontology," understood as a distinct branch of environmental philosophy.[18]

As for environmental technics, it is perhaps here that one may observe environmental philosophy's greatest blind spot. In keeping with the fact that philosophy of technology currently occupies a philosophical no-man's-land, and with the widespread view presented above that technology, as but a means to an end, is of no real concern to ethics, environmental philosophy has had relatively little to say about technology.[19] At a time of widespread consensus that an appropriate response to the environmental crisis requires an unprecedented technological revolution (e.g., renewable energy generation and storage; advances in energy efficiency; comprehensive recycling schemes; solutions to the pollution of air, water, and soils), environmental philosophers would appear rather naively to have assumed that, once we get our environmental ethics right, appropriate technologies will simply follow, in which case there is little or no need to develop a specifically environmental technics (or environmental philosophy of technology). And, lastly, while there has in recent years been a small number of attempts to develop the area of environmental epistemology, these remain marginal even within environmental philosophy, let alone with respect to philosophy as a whole.[20]

If biomimicry is an environmental philosophy that calls for a new ontology and that contains within it the foundations of a new technics, a new ethics, and a new epistemology, then it may help us overcome the current limitations of environmental philosophy. It may, in other words, allow environmental philosophy to develop beyond its starting point in environmental ethics, so as to cover environmental ontology, environmental technics, and environmental epistemology, while at the same time articulating and giving original content to these fields in a way that provides the philosophical foundations required to overcome the environmental crisis.

There is, however, a danger in seeing biomimicry as an environmental philosophy: environmental philosophy is of little or no interest to most contemporary philosophers. If philosophy of technology occupies a no-man's-land, environmental philosophy lives in the shadows, far removed from the philosophical limelight. Environmental ethics lives in the shadows of ethics, and the other branches of environmental philosophy, being little developed at all, remain veiled in almost total obscurity. The result is that the great existential crisis of our time—perhaps even the greatest crisis humanity will ever face—has for the most part been ignored by philosophers.

A personal anecdote illustrates this situation very clearly. I was once faced with the problem of how to categorize an environmental philosophy article of mine in an online philosophy archive. Eventually I managed to locate the closest category, environmental ethics, but only with great difficulty, for it had been classified as a subbranch of applied ethics, itself classified as a subbranch of ethics, which had in turn been classified as a subbranch of axiology (the study of value). At a time when we might reasonably expect of our deepest thinkers that they endeavor to lay the foundations for an appropriate response to an unprecedented existential crisis, that effort has been banished to a subbranch of a subbranch of a subbranch of axiology. The danger, then, in seeing biomimicry as an environmental philosophy is that it will remain in the shadows, unable to attract enough attention to bring about the philosophical revolution without which our attempts to resolve the ecological crisis are destined to remain superficial, and thus ineffective.

But, while this is indeed a danger, the way I propose to avoid it is not only to develop biomimicry in the sole context of environmental philosophy but also, and even primarily, *as a direct challenge to conventional philosophy*. Just as I earlier proposed a reworking of the basic *structure* of philosophy, so it is also important to propose an alternative to the basic *content* of philosophy. If environmental philosophy is to emerge from the shadows, it can only be by mounting a direct challenge to the conventional philosophy—with its near total obliviousness and indifference to the ecological crisis—that currently holds center stage.

## THE BIOMIMICRY REVOLUTION IN PHILOSOPHY

This brings me to the next topic of this introduction: the biomimicry revolution in philosophy. In responding to the ecological crisis, the biomimicry revolution aims for nothing less than the introduction of what Benyus calls a "new era" and thus a radical transformation of the world as a whole. This, I contend, requires a revolution in philosophy, for it is philosophy that thinks about the basic foundations of the world, and it is in revolutionizing these foundations that the world too may be revolutionized.

To understand the biomimicry revolution in philosophy, it is necessary to understand that with respect to which it constitutes a revolution. If biomimicry introduces an era in which nature is henceforth to be viewed as model, measure, and mentor, how is this era different from the present era, and indeed from other relevant eras prior to the present one? The answer, I suggest, is that these eras may also be characterized by the being (or beings) they take as model, measure, and mentor. These eras will be presented in greater detail at the outset of chapters 2, 3, and 4. In the meantime, I shall briefly summarize the principal results of these analyses.

In the Middle Ages, it was God who held the role of model, measure, and mentor. He was the ultimate source of all the models on which human designs were based, the standard against which right action was measured, and the spiritual guide to whom humanity turned for knowledge and wisdom. By analogy with biomimicry, we may characterize this medieval philosophy as "theomimicry."

With the advent of modernity, the role of model, measure, and mentor then fell to "Man." Human beings were taken as providing models for design, measures for ethics, but also mentors both for themselves and for one another, for they henceforth became the sole beings from whom it was possible to acquire knowledge and wisdom. Theomimicry, we may conclude, was displaced by "anthropomimicry."

More recently, in what some take to be the era of postmodernity, Man's occupation of the role of model, measure, and mentor has become doubtful, and it has even been suggested that any and every being may hold this role. What one might call "pantomimicry" has thus emerged as a rival to anthropomimicry. To the extent that we today are torn between the modern and the postmodern, it is between two different views regarding the rightful holder of the role of model, measure, and mentor—between anthropomimicry and pantomimicry—that we are torn.

The biomimicry revolution, then, is above all a revolution with respect to the being or beings we take as model, measure, and mentor: not God, not Man, not anything and everything, but nature. And yet, at the same time, we must also rethink what nature is, for we must ground the biomimicry revolution in a revolutionary understanding of that which is to hold the role of model, measure, and mentor. Nature can no longer be understood, as is generally the case today, simply as an object of scientific study about which philosophy, in contrast to science, has nothing in particular to say. The ancient discipline of ontology must be revived, and a revolutionary understanding of nature put forward.

As its etymology tells us, however, a revolution is not just something radically new, but a revolving back to something radically old.[21] And, in the case of the biomimicry revolution, that to which I believe we must revolve back is the very origin of Western thinking in ancient Greece, and thus also to that thinking that was eventually displaced by the theomimicry of the Middle Ages. It is to ancient Greece, and in particular to the early Greeks before Plato and Democritus, that we must look for a renewed understanding of the nature of nature—namely, the understanding of nature as physis, where physis refers to those beings that, in contrast to the artifacts produced by humans, *produce themselves.*[22] It is to ancient Greece that we must look for a renewed understanding of technics, and, more specifically, for a renewed understanding of technics (technē) as imitation (mimesis) of nature (physis). It is to ancient Greece that we must look for a renewed understanding of ethics—one grounded in the deepest sense of *ethos* as "inhabited place." And it is also to ancient Greece that we must look for a renewed understanding of epistemology, according to which knowledge (epistēmē) is one with, and not separate from, being (physis). It is, in short, to the creative renewal of four Greek words—*physis, technē, ethos,* and *epistēmē*—that we must look if we are to provide the biomimicry revolution with the philosophical foundations it requires.

But this revolving back to ancient Greece is not, and could not be, a simple repetition. It is not simply a question of transposing what the ancient Greeks said into our world, as if the revolution that awaits us could be accomplished simply by carrying out what the great German philosopher Martin Heidegger calls a "new onset of mere restoration and uncreative imitation."[23] Much important philosophical content remained only latent or was even completely absent from the thought of the ancient Greeks, entire sciences were still unknown, important theoretical and conceptual links had been missed, and great problems and obstacles had been introduced, especially by the time of Plato and Democritus. The biomimicry revolution, from this perspective, is neither something entirely new, nor an uncreative imitation of a distant past. Rather, it is a creative renewal of that past, one that does indeed reprise much of what the ancient Greeks had to say, and yet that also reveals things they did not say but that underpinned what they did say, that introduces elements or connections of which they were unaware, and that—perhaps most importantly of all—brings all of this together into an original composition.

## THE PRACTICAL SIGNIFICANCE OF THE BOOK

An important claim of this book is that it is only by seeing biomimicry as a new philosophy—one that addresses the deepest levels of our existence—that it will be able to offer a viable response, perhaps even the only truly viable response, to the ecological crisis.

My approach does, however, present a certain danger. More specifically, it presents the danger that all the talk of "revolution" remains but grandiose rhetoric, and that what follows is but a series of abstract discussions of little direct relevance to the urgent practical problems we face today. How, one might not unreasonably object, could reflecting on the language and thought of the ancient Greeks be of any help in resolving such concrete problems as climate change, resource depletion, and biodiversity loss? Surely only an ivory tower philosopher could possibly think that pondering the meaning of long-lost words like *physis, technē, ethos,* and *epistēmē* could have any bearing at all on the ecological emergency?

Whether or not this objection is valid depends at least in part on the power of philosophy, on what philosophy can do. It is certainly true that, over the last two hundred years or so, philosophy has been replaced by science as the primary locus of intellectual power. This began in earnest with Hume and Kant's rejection of metaphysics, and accelerated in the nineteenth and early twentieth centuries with the rise of the positivist belief that all genuine knowledge is either true by definition or arises ultimately from the application of reason and logic to data provided by sensory experience, as occurs in science. Contemporary philosophy—at least in the Anglo-Saxon world—may for the most part describe itself as "postpositivist," but it is still with respect to positivism that it defines itself.

There are some tentative signs that this situation may be changing. In particular, the rise of "object-oriented ontology" and various different "new materialisms" have led to a certain renewal of ontology.[24] No longer is nature the exclusive province of science, with philosophy being restricted to the realm of the human, for ontological reflection increasingly takes place with respect to natural objects. And yet, while this renewal of ontological engagement with nature is in some respects to be welcomed, it is insufficiently revolutionary. It leaves thinking divided between idealism and materialism, between subjects and objects, and simply to orient philosophical reflection away from human subjects toward material objects still leaves philosophy in a position of radical inferiority with respect to science, for in such a scenario philosophy typically concerns itself only with the *ontological implications* of nature as understood by science, not with developing its own understanding of nature.

To engage in a creative renewal of the thought of the ancient Greeks, by contrast, not only opens up the possibility of a new foundation for philosophy but also taps into a time when philosophy did possess genuine power, when it was not secondary to science, but, on the contrary, provided the ground from which the sciences could spring forth. If, then, it is today necessary creatively to renew the thinking of the ancient Greeks, as opposed to simply sharpening, polishing, and being dazzled by philosophy's cutting edge, it is precisely because, since it was there that the first thinking of being, making, acting, and knowing first began, it is also there that we must seek what Heidegger calls a "new beginning."

Returning, then, to the practical significance of the present work, and in particular its relevance to the ecological crisis, there are two key claims I wish to make: first, the deep drivers of the ecological crisis are philosophical in nature, or at least can only be adequately understood and dealt with through philosophy; second, these deep drivers cannot be adequately understood and dealt with by tinkering with or even advancing philosophy as we currently find it, for that will leave them largely intact, but only by returning to the very origins of philosophy and seeking out a "new beginning" for Western thought. From this perspective, to think that the deep drivers of the ecological crisis lie in such things as burning fossil fuels, urbanization, industrial agriculture, and population growth is mistaken. These are drivers of the ecological crisis, but they are superficial ones, and as long as the deep drivers remain intact, the superficial ones will remain intractable.

The danger remains, however, that, in addressing the deep drivers of the ecological crisis by means of philosophy, their connection to the superficial drivers will remain veiled. If the biomimicry movement has thus far been overly practical in orientation, this book presents the converse danger of being overly theoretical in orientation, or at least of not making sufficiently visible the connections between theory and practice. To some extent, I embrace this theoretical orientation. It would fall outside the book's scope *both* to spell out the philosophical foundations of biomimicry *and* to apply the philosophy thus set out to the myriad practical problems we face today. But this is not to say that the present work is altogether devoid of practical significance. Chapter 1 may be highly theoretical, advancing as it does a theory of nature as physis or self-production, and so not dealing directly with biomimicry itself, but chapters 2, 3, and 4 are of direct relevance to practical matters.

The theory of biomimetic technics advanced in chapter 2 consists primarily in the setting out of a comprehensive framework for understanding the process of biomimetic innovation. This framework deals with many issues familiar to anyone working in biomimicry research and development, from the question of the different types of traits one may abstract from nature for the purposes of imitation to the vexed issue of the relation between imitation and inspiration. In so doing, it also presents and discusses many concrete instances of biomimetic innovation, from brain-inspired computing to analog forestry.

Chapter 3 situates biomimetic ethics within a new approach to environmental ethics I call "environmental action ethics," which, in focusing on concrete actions, especially preservation, conservation, restoration, and imitation, has the potential to make environmental ethics much more relevant and accessible to practitioners and decision-makers than is currently the case. As for the theory of biomimetic ethics I go on to propound, while it is grounded in the abstract and general principle of measuring ourselves against the ecological standard set by Gaia, when applied to specific practical contexts it gives rise to quantitative objectives against which the "rightness" of artificial products and systems may be measured.

As for chapter 4, in laying out the theoretical basis of biomimetic epistemology, it not only establishes potentially important links with other epistemologies, including feminist and indigenous epistemologies, but it also lays the groundwork for the theory of "biomimetic science" advanced in the second half of the chapter. This latter theory in turn provides a clear vision of what counts as progress in biomimetic innovation; far from being simply an innovation strategy one may or may not adopt, biomimicry is thus equipped with a "scientific method" capable of delivering measurable progress.

In view of all this, it is clear that this book sets out the philosophical foundations of biomimicry, including their relation to other philosophical and theoretical movements, from environmental ethics to feminist epistemology, while also seeking to underpin, orient, and contribute to more practically oriented work taking place—or that may in the future take place—both in biomimicry research and in the broader environmental movement. Nevertheless, since it concentrates on the philosophical foundations of biomimicry, it leaves ample scope for more practically oriented readers to work out the practical implications and ramifications of this underlying philosophy, just as it leaves ample scope for more theoretically oriented readers to analyze, critique, or develop the underlying philosophy itself.

## CHAPTER 1

# *Nature as* Physis

## An Ontology for Biomimicry

The three principles Benyus sets out at the beginning of *Biomimicry* have one obvious thing in common: the word "nature." More specifically, they call on us to embrace three different—but presumably complementary—ways of viewing nature: as model, as measure, and as mentor. But what is this thing, nature, that biomimicry calls on us to view in these three different ways? Without an understanding of what nature is, we cannot understand what we are supposed to be viewing as model, measure, and mentor, in which case biomimicry cannot even begin. And yet, if these principles require us first to inquire into what nature is, this is not to say that our inquiry should proceed independently of all awareness of them; they clearly provide constraints on how we might answer the question of the nature of nature. Nature must be positively identified and understood in its own right if it is then to be viewed as a model, measure, and mentor; and yet that characterization must be of something that one could take as model, measure, and mentor. So, with this dual task in mind, let us turn our attention to the question of the nature of nature.

The meaning of the word "nature" is famously inconstant and polysemic. It has varied significantly between cultures, periods of history, and schools of thought. Even within everyday contemporary usage, it is subject to great variation, depending on context. What it signifies in the expression "the laws of nature" is quite different, for example, from what it signifies in the expression "the protection of nature." Noting this, it has sometimes been claimed that we would be better off avoiding the word altogether, preferring more precise terms adapted to the specific contexts in question. For example, one could talk instead about the "laws of physical reality" and the "protection of biodiversity."

It has also been noted that many languages lack any direct equivalent to our word "nature"—an observation that has in turn given rise to and supported the claim that the word and concept of nature is specific to Western civilization.[1] At a time when the globalized West is doing unprecedented environmental damage, it has even been suggested that the word and concept of nature is to blame, that the solution to the environmental crisis is not to save or protect nature but, paradoxical as it may seem, to do away with it altogether. As Bruno Latour writes: "When the most frenetic of ecologists cry out in fear: 'nature is going to die,' they are unaware how right they are. Thank God, nature is going to die. Yes, the great Pan is dead! After the death of God and of Man, nature too must come to an end."[2]

And yet, even if it is true that the word "nature" is inconstant and polysemic—and specifically Western—it is far from clear that this means we should advocate and celebrate its demise. Its variability could be rather a sign of a profound fertility and plasticity, of an exceptional capacity for semantic renewal lacking from more precise contemporary substitutes or alternatives. Likewise, while the word and concept of nature would indeed appear to be specific to, and perhaps even foundational for, the Western tradition, this realization could lead to the conclusion that we need a better understanding of the place and role of "nature" in that tradition, so that we may become more fully aware both of its history and of any viable semantic transformation or renewal that may be open to it in the future. This is not to deny that contemporary understandings of "nature" are problematic, but to assert the word and concept of nature should be abandoned altogether does not follow.

This brings me back to the use of the word "nature" in biomimicry. Biomimicry is a revolutionary proposal for responding to the ecological crisis, and, at least in its initial formulation by Benyus, it makes foundational use of the word "nature," both in the statement of its three basic principles and at other critical moments, such as when Benyus describes what she calls "nature's laws, strategies, and principles."[3] To abandon the word "nature," it would seem, would also be to abandon biomimicry.

In response to this, it could no doubt be objected that one could reformulate both the basic principles and the general discourse of biomimicry in such a way that the word and concept of nature would in every case be expunged. But, given the power of "nature" to participate in the generation of such a radical new approach to the ecological crisis as biomimicry, it is far from clear that we would be better off seeking to avoid or eliminate it. A less extreme but still radical approach—one that I shall adopt in this chapter—is to keep the word "nature" but to interpret it in a revolutionary manner. The biomimicry revolution, from this perspective, would be not only a revolution in our *relation* to nature but also in our *understanding* of nature. This, then, shall be the objective of the present chapter: to put forward, develop, and explore a revolutionary understanding of nature. But, rather than leap directly into this task, let us consider two obvious, but ultimately problematic, ways one might attempt to interpret the word "nature" in the context of biomimicry.

The first of these consists in understanding the word "nature" in the manner of scientific naturalism. Scientific naturalism has two components: first, an ontological component, according to which nature is identified with being. From this perspective, anything that truly is belongs to nature and, as such, is natural. An important implication of this is that there is nothing supernatural, nothing that in some sense goes beyond nature, that is not a part of nature. The second component is methodological and epistemological; it is science alone that allows us to gain knowledge of nature, with the important implication that philosophy cannot engage in enquiry into nature using extrascientific methods. To the extent that philosophy has anything to contribute to the thinking of nature, it is not as something essentially different from science, but as an extension or continuation of science in its more abstract conceptual and theoretical aspects.

Putting together these two components of scientific naturalism, one arrives at the following idea: that nature is what science says it is—namely, the totality of elementary particles, the forces governing their interactions, and the phenomena that result from the effects of these forces on those particles at all different levels and scales. Viewed in this manner, nature clearly includes the entities and phenomena studied by the natural sciences, notably physics, chemistry, and biology. As for humans, they, too, are considered part of nature and may be studied using the methods of science, such that any characteristics specific to humans must be seen as arising ultimately from the effects of the fundamental forces on elementary particles at higher levels of organization. The various dualisms holding that humans possess something (e.g., a soul, a transcendental self) that makes them radically different from the rest of nature, and that is not open to scientific investigation, are thus rejected.

In the context of the present inquiry into biomimicry, there is an obvious problem with the view of nature characteristic of scientific naturalism: it is far too broad. Science may study and describe a great many things that it would make no sense whatsoever to imitate, let alone to take as an ethical standard or a source of knowledge. We live in a universe that is—with the notable exception of our own planet—largely inhospitable to life, and imitating the vast majority of astrophysical phenomena would be at best pointless or impossible, and at worst profoundly dangerous. Our neighboring planet Venus, for example, may well have undergone runaway climate change—something we wish to avoid, not reproduce.

In response to this, it could perhaps be objected that the imitation of nature is necessarily selective, that we have no obligation to imitate each and every natural entity or phenomenon, and that the view of nature characteristic of scientific naturalism has the advantage of providing the broadest possible set of entities and phenomena from which we may then carefully select appropriate models, measures, and mentors. But to conceptualize nature in that way nevertheless remains unhelpfully broad; only a subset of the beings or phenomena described by natural science could ever be taken as models, measures, and mentors, in which case an understanding of nature would appear to be required that corresponds to that subset, not to the total set from which the subset is drawn.

This brings me to the second obvious way that nature might be characterized: as life. From this perspective, the subset of phenomena to which biomimicry calls on us to turn in search of models, measures, and mentors are the living ones studied by biologists. This view is certainly not incompatible with scientific naturalism; one could uphold scientific naturalism, understood as an ontological doctrine, but then affirm, in keeping with the famous polysemy of the word "nature," that there is a second, narrower sense of the word—nature as life—operative in the context of biomimicry. In support of this interpretation, one could point both to the etymology of the word "biomimicry," which would appear to tell us that what is to be imitated is "life" (*bios* in Greek), but also to the fact that this interpretation corresponds quite closely to how biomimicry is generally understood and practiced today.

But, while this biological interpretation of nature is clearly preferable, at least in the context of biomimicry, it is not without problems. One problem is that it raises the question of why anyone—and Benyus in particular—would refer specifically to nature in the context of biomimicry. If nature means the same thing as life, why not just talk directly of life? Is the use of the word "nature" simply an ethnocentric habit that would ideally have been overcome by using the more precise word "life," or is there something about the word "nature" that, on closer inspection, actually makes it more appropriate in this context, in which case it would be a mistake, even here, simply to identify nature with life?

A second problem is that to interpret nature as life does not answer, but instead displaces, the basic question at hand; the question is no longer what *nature* is but rather what *life* is. To identify nature with life may narrow the field of inquiry from being in general to a certain region of being, the living, but the question of what defines that region would still need to be answered. This is not to say that an answer could not be forthcoming, but simply that it might present major challenges—perhaps even insurmountable difficulties—both in itself and as far as the role of the concept of life within biomimicry is concerned.

A third problem, which follows on from the problem of appropriately conceptualizing life, is that it is far from clear what entities are to count as life or as living. Biological organisms clearly do, but what about other entities studied by biologists, such as societies, species, and populations of living beings? What about entities studied by ecologists, such as ecosystems? Do they count, and, if not, should we see the imitation of ecosystems as something quite different from biomimicry—namely, ecomimicry?

In response to these difficulties, it may seem that it would be better to talk here of living systems, where the concept of a living system is understood to include not only biological systems, like living beings, societies, or species, but also ecological systems. However, this still leaves open the question of what a living system is—that is to say, what semantic unity the concept has, such that it may meaningfully describe, rather than simply designate by stipulation, such different entities as organisms, societies, species, and ecosystems. After all, one may easily stipulate that the phrase "living system" applies to many biological and ecological phenomena, from organisms to ecosystems, but it isn't obvious what common properties these phenomena share such that it makes sense to group them together under this particular concept. To say, for instance, that a living system is a system that is *either* a living being *or* that includes living beings as an essential component, as in the case of societies, species, and ecosystems, may well describe how the expression "living system" is typically used today, but it leaves open the question of what the two sides of this disjunctive definition have in common—what basic unity holds them together—so that it makes conceptual sense to designate them both by the same phrase.

Lastly, a fourth problem is that we may wish to imitate certain phenomena that we would not hesitate to describe as "natural," but that, more so even than ecosystems, would not usually be considered living or even to belong to the class of living systems. In *The Shark's Paintbrush: Biomimicry and How Nature Is Inspiring Innovation*, Jay Harman describes how he invented a new water mixer by imitating the form of a vortex.[4] But few people, if any, would consider vortices to be alive or to be living systems. So should we expand our understanding of biomimicry so that it extends to the imitation of certain nonliving systems, which would present the problem of how to define nature in a way that was broader than life but narrower than the totality of physical reality? Or should we limit our interpretation of nature to the living such that the imitation of nonliving phenomena, such as vortices, would not count as biomimicry? The word "biomimicry" implies the latter, but that simply raises the question of whether our investigation should be fundamentally constrained by the contemporary neologism "biomimicry," or whether the real object of our inquiry is something else, the imitation of nature, with "biomimicry" being but the most appropriate—yet still far from perfect—way of describing and labeling that object with a single word.

Faced with such labyrinthine problems, there are no doubt a great many paths that one might look to explore, and there can be little doubt that an important part of philosophical inquiry into biomimicry is to investigate, evaluate, and compare their respective viability. But, instead of undertaking that exercise here, I shall simply present the path that I have taken in my own thinking about nature, understood both as a topic of philosophical reflection and as a foundational element of biomimicry. This path does not involve taking a readily available understanding of nature—as the totality of entities and phenomena that may be studied by science, or as life or living systems—but, rather, journeying back to the very origin of Western thinking about nature so as to renew and revive the early Greek concept of physis.

## NATURE AS PHYSIS

### *The Concealment of Physis*

The ancient Greeks spoke of *physis*, not of "nature"—a word of Latin origin (*natura*) used by the Romans to translate *physis*. But what was or what is *physis*? Consider the following characterizations put forward by the twentieth-century German philosopher Martin Heidegger:

We shall now translate *physis* more clearly and closer to the originally intended sense not so much by growth, but by the "<u>self-forming</u> prevailing of beings as a whole."[5]

*Phuein* means to let grow, procreate, engender, produce, primarily to <u>produce its own self</u>.[6]

The being of what <u>emerges and comes to presence on its own</u> is called *physis*.[7]

*Physis* [is] the <u>arising of something from out of itself</u>.[8]

Obscure, vague, and decontextualized as these short quotations may be, they do—as the underlined sections highlight—seem to converge around a common idea: that *physis* names that which in some sense or other "produces itself" or "brings itself to presence." This is not to say that when the ancient Greeks sought to inquire into physis they undertook a direct inquiry into the meaning or nature of this self-production, self-presenting, or, to use another Greek expression employed by Heidegger, *poiēsis en heautōi*.[9] On the contrary, the concept of self-production or self-presenting remained latent in their thought, with physis being instead identified with some specific entity or other. For Thales, physis was water. For, Anaximenes, it was air. For Heraclitus, it was fire. And so on and so forth.

It is not so much that the entities with which physis was identified were not thought in the light of the basic concept of physis—in other words, as in some sense or other "arising on their own," rather than as just different types of matter, as Aristotle thought.[10] But the concept of physis itself, considered prior to its identification with some entity or other, was not a direct object of inquiry. If we today, however, are to understand physis itself, we cannot proceed in this manner, but must instead undertake a task that even the early Greeks never carried out: to understand physis itself, to understand the nature of self-production or self-presentation prior to identifying it with any particular entity or set of entities.

Rather than analyze the concept of physis directly, let us first consider how it was that early Greek thinking of physis was rejected and replaced by other ways of thinking. It is common practice to divide the history of Greek philosophy into two periods: the prephilosophical period of the pre-Socratics, which occurred prior to Socrates and Plato, and the period of philosophy proper, which began with Socrates and Plato and continued through to Aristotle.

Heidegger famously rejected the term "pre-Socratics" on the grounds that it reduces those it names to mere precursors of Socrates and Plato, and therewith also to philosophy proper. He himself refers to them instead as "early Greeks," for this expression allows them to be seen as important thinkers in their own right—though not, strictly speaking, as philosophers, for the word and concept of philosophy arose only with Plato.[11] There is nevertheless one thing that Heidegger shares with mainstream thinking about this period: the idea that a single basic rupture took place with Plato. Prior to Plato, there were early Greek thinkers of physis; after Plato, there was philosophy proper.

Where I depart from Heidegger is in my belief that the fundamental rupture with early Greek thinking of physis occurred not with Plato, but with Parmenides. It was Parmenides, not Plato, who first rejected early Greek thinking of being as physis. Why did he do so? With a view to answering this question, let us concentrate on a specific claim in fragment B8 of Parmenides's poem: that "what is" (*esti*) cannot come from nothing.[12] There are, Parmenides thinks, two reasons for this. First, to think that nothing existed prior to what is is to think that nothing can be, but nothing is precisely what is not; it doesn't exist. Second, even if nothing did exist prior to what is, there is no reason that what is should have emerged from it. There is, in other words, nothing in nothing that could have produced something.[13] What is, Parmenides concludes, cannot have come to be, and, for essentially the same reasons, could never pass away. But coming into being and passing away were fundamental traits of physis. Consider the following words of Anaximander, which, as Heidegger reminds us, belong to the oldest extant fragment of Western thought: "Whence things have their coming into being there they must also perish."[14] Things, Anaximander tells us here, come into being and pass away, which is precisely what Parmenides denies, and, in denying this, Parmenides also rejects the very concept of physis.

In rejecting physis, Parmenides laid the basic ontological ground for all subsequent Western thought. In order to see why, we must first consider a recent challenge to the orthodox interpretation of his work. It is generally supposed that Parmenides was a numerical monist, that he thought that what is is one in the numerical sense—or, put simply, that there is only one thing.[15] In recent years, this has been called into question. Building on the work of Alexander Mourelatos,[16] Patricia Curd has argued that Parmenides's aim was not to expound and justify numerical monism, but rather to work out the properties that any true being must possess, without assuming there to be only one of them.[17]

Three such properties are of particular significance here. The first property, which we have already looked at, is that any true being can neither come to be nor pass away. The second is that any true being cannot change. And the third is that any true being cannot appear to the senses, only to the mind. The question, then, for Parmenides's successors, was which beings possess these three properties. Two answers emerged simultaneously: the Ideas (with a capital I), which Plato thought were eternal, unchanging, and invisible to the senses; and the Atoms (with a capital A), which Democritus thought possessed these same three characteristics.

Combined with Parmenides's rejection of physis, Curd's reading of Parmenides suggests an alternative model of ancient Greek thought to that which currently prevails. Whereas standard classifications of ancient Greek thinkers assume a linear model according to which everyone is either a pre-Socratic or a post-Socratic (see fig. 1.1) and so typically see Democritus as just another of the former, according to this alternative model (see fig. 1.2) a rupture occurs with Parmenides, and thinking then bifurcates into, on the one hand, Platonic idealism, and, on the other, Democritean atomism. But this bifurcation should not, strictly speaking, be seen as a bifurcation *within philosophy*; on the contrary, what Parminedean ontology made possible was the bifurcation of thinking not only into idealism and atomism but also—and more profoundly—into philosophy and science.

That philosophy emerged with Plato is visible both in the fact that it was Plato who coined the word "philosophy," and in that the basic aim and method of philosophy also goes back to him.[18] Philosophy, as conceived by Plato, is metaphysics, understood as the attempt to use pure reason to gain access to a realm located beyond the physical. This is not to say, of course, that subsequent philosophy has embraced the details of Plato's philosophical system, but simply that all

Pre-Socratics ⟶ Plato ⟶ Philosophy as footnotes to Plato

**1.1** The linear model of Western thought

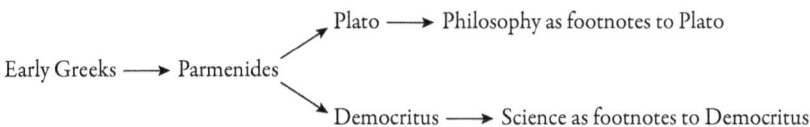

Early Greeks ⟶ Parmenides ⟵ Plato ⟶ Philosophy as footnotes to Plato
Democritus ⟶ Science as footnotes to Democritus

**1.2** The bifurcation model of Western thought

subsequent philosophy—bearing in mind that philosophy proper only existed subsequent to him—is working within a tradition that began with Plato. Kant's noumenal realm of "things in themselves," for example, may be quite different from Plato's realm of Ideas, and yet it shares with Plato's realm the common traits of being both radically separate from the physical sphere, and, to the extent that we may access it at all, we require pure reason to do so.[19] Philosophy, we may conclude with Alfred Whitehead, is indeed "footnotes to Plato."[20]

That science emerged with Democritus is less immediately apparent. This, I think, is due in large part to the traditional classification of Democritus as a pre-Socratic and, implicitly, as someone whose thought was superseded by Plato. But Democritus's thought was not superseded by Plato. The basic approach Democritus advocated, which involved cutting reality up into its fundamental units (the "A-toms," from the privative *a-* and *temnein*, meaning "to cut"), and then working out how it is that they interact with one another so as to form composite entities, has been fundamental to all subsequent science, including that of our own time. Indeed, the profound connection between atomism and science is visible in the very etymology of the word "science." The root meaning of "science" is not "knowing," as one might suppose, but "cutting," as is still visible in the contemporary French word *scier*, meaning "to saw."[21] But there is a limit to how far "science" can go: the A-tom, that which cannot be cut.[22] And so, having cut reality up into its most basic units—the elementary particles of the standard model, for example—the only remaining task is to work out how and why it is that they interact, what phenomena result from these interactions, and how and why they do so. If philosophy is footnotes to Plato, science is footnotes to Democritus.

Of these footnotes, two are particularly important. The first footnote is that modern science and philosophy differ from the science and philosophy of the ancients in that they believe in the existence of forces or mechanisms that govern the interactions of, in the first case, the Ideas, and, in the second case, the Atoms. Whereas Democritus thought that the Atoms traveled through the void, interacting only on account of their shape, modern science has shown that their interactions are governed by four fundamental forces: gravity, electromagnetism, and the strong and weak nuclear forces. And whereas Plato thought that the Ideas exist independently of one another in some sort of extraworldly realm, modern philosophy locates the ideas in the human mind and then asks what basic mechanisms there are (e.g., Hume's association of ideas,[23] Kant's syntheses[24]) that explain their interactions with one another.

The second important footnote is that the Parmenidean ontology of "permanent enduring," which first allowed philosophy and science to begin, has been largely abandoned. Ideas, it is now widely assumed, may come into being and pass away, and they may also undergo change.[25] The same goes for Atoms (i.e., elementary particles), which, as contemporary physics has shown, may not only pass into and out of existence but also both change into one another and take on the character of a wave, provided that the principle of the conservation of energy is respected. The only one of Parmenides's criteria for true beings that would appear still to hold in modern science and philosophy is the one that says that they cannot appear to the senses, but only to the mind, for no one has ever seen an idea or an elementary particle (though their presence can be detected through the traces they leave behind). But, while we may live at a time when, to use an oft-heard slogan, becoming has triumphed over being (where being is understood in the Parmenidean sense of "permanent enduring"), the basic division of the world into Atoms and Ideas, matter and mind, science and philosophy, still reigns.

It is also true, however, that in the long-standing battle between materialism and idealism, between science and philosophy, science has emerged triumphant. With the rise of scientific naturalism, philosophy is no longer seen as providing a viable alternative to science. To the extent that philosophy *practices* scientific naturalism, it limits itself to being but a part of science and thus assumes the Lockean status of a handmaiden to science; and to the extent that it *advocates* scientific naturalism, it renounces itself, assuming instead the status of a cheerleader for science, a supportive sideshow offering only encouragement to teams of scientists in their attempt to obtain ever more detailed and comprehensive knowledge of nature.

But, when this is how things stand, what happens to those topics that, under the previous division of labor between philosophy and science, were seen as the province of philosophy, especially those related to the soul, viewed as both a metaphysical entity and a moral agent? They are increasingly covered by systems theory and cybernetics. If everything is but Atoms and composite systems of Atoms, the task we face is to understand the workings of those composite systems, including the various mechanisms that regulate or govern them. In keeping with this, Norbert Wiener, in his 1948 book *Cybernetics*, applied the basic concepts of cybernetics first to animals and machines, and then, in his 1950 book, *Cybernetics and Society*, to the human mind, society, language, law, and so on.[26] The result was that those areas of inquiry that had previously seemed so distant and different from the Atoms and their composite formations that they could not readily be explained by science, only by philosophy, could be explained in terms of the regulatory mechanisms of complex systems. Ethics, for example, was no longer a separate domain requiring philosophical investigation into what Plato saw as the highest Idea of all, the good, or, as in Kant, into the limits imposed by reason on the freedom of the transcendental subject, but rather a set of norms and beliefs responsible for the self-regulation of social systems. This, then, is why Heidegger claimed that, with the advent of cybernetics, philosophy has come to an "end."[27]

And yet, according to Heidegger, hidden within the "roots" of cybernetics, there lies the possibility of a radical return to early Greek thinking of physis.[28] And this, as we shall now see, is precisely what happened. The basic trait shared by the complex systems studied by Wiener was self-regulation. In *Cybernetics*, for example, the concept of self-regulation was applied both to the new generation of cybernetic machines and to animals. Just as a boiler may regulate its temperature on its own—that is to say, through first detecting and then correcting any deviation from a target temperature—so any deviation from the target temperature of a mammal will be first detected and then corrected through such mechanisms as sweating or shivering. Both living beings and machines, both nature and technology, may thus be described using the same basic conceptual and theoretical framework, in terms of the information, communication, and feedback involved in self-regulation.

What later became apparent, however, was that, in collapsing the distinction between natural and technological entities, both of which were henceforth conceived as "self-regulating systems," cybernetics overlooked the very different ways in which natural and technological entities *come into being*. Technological entities are produced by human beings, but that is not the case for natural ones. So, how are they produced? The revolutionary answer that emerged—and that takes us right back to the characterizations of physis with which we began—was that they *produce themselves*; that they are, in a word, "self-producing."

## The Rediscovery of Physis

The concept of self-production is today associated primarily with the Chilean biologists Humberto Maturana and Francisco Varela, who first theorized it in the 1970s.[29] Maturana and Varela claim that the fundamental trait of living beings, and also what sets them apart from technological entities, is self-production—or, as they call it, "autopoiesis." The application of cybernetic concepts to biology, they conclude, is fundamentally problematic, for it obscures the fundamental difference between life and technology: whereas technology is produced by humans, living beings produce themselves.

But, before looking directly at the concept of autopoiesis, let us first note a number of limitations with Maturana and Varela's understanding of this concept. The first is the naivete present in their discussion of the concept's originality. According to Maturana, autopoiesis is a "word without a history," coined by him to describe the basic trait of the living.[30] However, when one recalls Heidegger's many analyses of the Greek understanding of physis in terms of "self-" (auto-) "production" (poiēsis), this claim to absolute originality starts to appear doubtful. Rather than having invented an entirely new concept, Maturana and Varela would appear instead to have unknowingly rediscovered and appropriated one that remained latent in the thought of the early Greeks: physis. Further, it is not difficult to see why they may have failed to see any connection between autopoiesis and physis. This latter concept was so completely suppressed by the post-Parmenidean traditions of both philosophy and science that, when it finally reemerged—exactly where Heidegger thought it would, in the roots of cybernetics—it was believed to be a pure invention.

The second limitation is that Maturana and Varela identify autopoiesis with "life" or the "living." Quite why they do this is not entirely clear, but it may be a product of their disciplinary specialization. As biologists, when they realized that living beings were self-producing or autopoietic, they presumed to have discovered the essential trait of living beings rather than a much broader phenomenon of which living beings are but one instance. Further, since they were writing at a time when early Greek thinking of physis had long since been rejected and forgotten, there was nothing other than life with which they could readily identify autopoiesis. Lacking any conceptual alternative, they identified autopoiesis with life, the closest contemporary equivalent to physis.

The third limitation is a consequence of this identification of autopoiesis with life. The ancient Greeks, Heidegger tells us, saw physis as "being itself."[31] The thinking of physis, it follows, was not simply an inquiry into "nature," but into being itself. By contrast, since Maturana and Varela identify autopoiesis with life, their thinking about autopoiesis lacks the ontological awareness and significance of early Greek thinking of physis. The nature of living beings may, on Maturana and Varela's account, lie in the trait of self-production, and yet the extraordinary ontological significance of this trait goes largely unnoticed. The upshot is that, while Maturana and Varela's thinking about autopoiesis appears highly relevant to the philosophy of biology—even to the philosophy of mind and epistemology—its relevance to ontology has been largely overlooked.[32]

If the concept of autopoiesis had remained restricted to the thought of Maturana and Varela, there would still be much work to do to articulate autopoiesis with ancient Greek thinking of physis. Importantly, however, the contemporary French thinker Edgar Morin has already accomplished a significant part of this task, especially in his 1977 book, *La méthode, tome 1: la nature de la nature*.[33] As its subtitle suggests, this work is much more ambitious in scope than that of Maturana and Varela, for its aim is much the same as that of this chapter: to answer the question of the nature of nature. Morin's answer is that nature is "self-production," a concept he arrives at in a similar manner to Maturana and Varela: through a critique of cybernetics and systems theory.[34] Cybernetics, Morin tells us, studies processes of communication, feedback, and so on, in self-regulating systems, but it fails to address the question of how the systems in question come into being in the first place. His answer is that, whereas in the case of technology, it is *we humans who produce them,* in the case of nature, they *produce themselves.*[35]

Morin further claims that the concept of self-production is what the early Greeks meant by physis.[36] Quite why he says this is not entirely clear; he at no point seeks to explain this alleged identity between physis and self-production in a direct and comprehensive manner. But the very fact that he does identify his understanding of nature in terms of self-production with the Greek concept of physis nevertheless reinforces this chapter's basic thesis: that it is possible to arrive at a revolutionary interpretation of nature through making explicit and clarifying something that remained implicit and obscure in early Greek thinking of physis: the concept of self-production.

In keeping with Morin's identification of self-production with nature—or, more precisely, with physis—his thought also differs from that of Maturana and Varela inasmuch as he does not identify self-production with life, the living, or even with living systems. On the contrary, Morin extends the concept of physis both into the realm of the physical, arguing that many purely physical beings or systems, including, notably, atoms (in the contemporary nonetymological sense), vortices, and stars, are self-producing,[37] but also into the realm of the social, arguing that human societies are self-producing[38]—an idea developed also by the German sociologist Niklas Luhman on the basis of Maturana and Varela's concept of autopoiesis.[39] Thus it is that physis, for Morin, assumes the status of a transdisciplinary concept present across, but also foundational for, all the sciences, from physics to sociology.[40]

Lastly, Morin is aware of the ontological significance of the concept of self-production. One of the fundamental limitations of systems theory, he tells us, is that, since its basic focus is on systems rather than beings, it is not concerned with questions of fundamental ontology, including such long-standing questions as the being and existence of beings.[41] Further, since Morin explicitly defines production as "bringing into being and/or existence," it follows that beings that self-produce possess a trait that, viewed ontologically, is indeed quite extraordinary: *they bring themselves into existence.*[42] And yet it is also true that Morin does not put forward anything like a clear and comprehensive ontology based on the concept of self-production; his reflections on such key concepts as being and existence are scattered throughout his text and only rarely linked to the work of other philosophers or philosophical traditions.

Another thinker who made a major contribution to the rediscovery of physis, particularly in its ontological aspects, is Heidegger. As we have already seen, Heidegger claimed that the thinking of "being itself" first emerged in early Greek thinking of physis. And yet he also thought that being itself became forgotten or concealed—something that I think began not with Plato, as Heidegger himself thought, but with Parmenides. But, apart from providing the basis of a historical narrative of the forgetting or concealment of physis, what contribution has Heidegger made to the thinking of the concept? With a view to answering this question, let us note that whereas Morin's thinking about physis arose primarily from a critical engagement with cybernetics and systems theory, Heidegger's came primarily from a critical engagement with phenomenology, the study of that which appears. Despite the critical aspects of these engagements, the traditions with respect to which they occurred continued to exert a major influence on each thinker. In the case of Morin, the enduring influence of cybernetics and systems theory led him to focus on the concrete processes or mechanisms involved in self-production. In the case of Heidegger, the enduring influence of phenomenology led him to focus not only on self-production but also on self-bringing-forth or self-bringing-to-presence.

This difference is particularly manifest in the two thinkers' quite different ways of understanding poiēsis. For Morin, this word signifies producing in the sense of making or creating.[43] For Heidegger, it signifies producing both in the sense of making or creating (*herstellen* in German) and in the sense of bringing forth or presenting (*darstellen* in German)—a sense manifest also in such expressions as "to produce something from one's pocket" or "to produce a witness."[44] From this latter perspective, *physis*, understood as *poiēsis en heautōi*, means not only self-producing in the sense of self-creating but also in the sense of self-presenting. Any being that belongs to physis, it follows, does not simply produce itself in Morin's sense, for it also, as Heidegger at one point puts it, "places itself into appearance."[45]

This phenomenological aspect of physis in turn ties in with a key concept of Heidegger's that has no counterpart whatsoever in the thought of Morin: the "clearing" (*die Lichtung*), understood as the space where beings in one way or another come to presence.[46] Physis, from this perspective, is the primary but not the only way that beings may come to presence in the clearing. Heidegger's principal contribution to the thinking of physis, it follows, is to have provided this concept with a phenomenological dimension while introducing a closely related concept—the clearing—that, as will become apparent later on, may play a vital role in the theorization of a new ontological framework encompassing also the human relationship to nature.

## The Concept of Physis

In the previous section, we looked briefly at a series of thinkers, all of whom have made major contributions to the rediscovery of physis: Maturana and Varela, Morin, and Heidegger. What we have yet do is analyze this concept, such that the nature of self-production may be more clearly understood. To this end, I shall draw in what follows on the thinkers presented above with a view to modifying, developing, and synthesizing some of their key insights in such a way that an original understanding of the concept of physis may emerge. Since this work of modification, development, and synthesis is rather complex, however, I shall not engage in a critical discussion of the thinking of self-production carried out by my forebears and how it differs from my own. Instead, I shall simply present my own interpretation of physis, which, as I understand the concept, has three components: 1) self-construction; 2) self-delimitation; and 3) self-bringing-forth.

1. Self-construction

A being that "constructs itself" is a being whose parts are not produced by some other being, but by other parts of itself. This production may be either linear, as is the case when some parts produce other parts (As produce Bs), or it may be circular, as is the case when some parts produce other parts, which in turn produce the first parts (As produce Bs; Bs produce As). In both cases, however, the construction of the entity remains *internal* to it, such that the being in question may be said to construct itself through the production of its own parts (by other parts of itself).

2. Self-delimitation

A being that "delimits itself" is a being that produces a limit or boundary that sets it apart from what thereby becomes its environment and thus other than itself. This limit or boundary may either be a specific part of the entity in question, or it may be an emergent phenomenon arising from the way in which the entity's parts produce its other parts (i.e., from the way it constructs itself). This limit or boundary is not, however, simply the point at which the thing ends; rather, it in one way or another holds the parts of the thing together, such that they come to form a finite entity, distinct from the rest of reality.[47]

3. Self-bringing-forth

A being that "brings itself forth" is a being that places itself into appearance and, as such, is responsible for its own coming to presence. The way that it does this is through constructing and delimiting itself. And yet self-construction and self-delimitation are insufficient for a being to bring itself forth; that requires a "space of appearance" or "clearing" into which it may bring itself, thus becoming open to those beings that are, as it were, exposed to the clearing, in the sense of being open to any entity present therein. Further, were the entity not to appear in the clearing, it would be impossible for any beings exposed to the clearing to categorize it as "self-producing." It is only because it is "there" that they may say of it that it is self-producing, and, conversely, it is only because it produces itself—where to produce itself is understood in a broad sense inclusive of self-bringing-forth—that it is "there."

---

Self-production, as the above analyses make clear, has three components. It is important to realize, however, that these components cannot exist independently of one another, and it would thus be impossible for one of them to occur without the others, or even two of the three without the other one.

Self-construction arises through the production of the parts of an entity by its other parts. But, as we have seen, parts can only be parts by virtue of being located within an entity that is delimited from its environment. In the case of self-producing entities, this limit must be produced by the entity itself, such that we may say of the entity that it "delimits itself." Self-construction, therefore, depends on self-delimitation. Conversely, an entity could not delimit itself other than through the production of a limit or boundary that is either a part produced by other parts, or an emergent phenomenon arising from the production of parts by other parts. Self-delimitation is also dependent on self-construction.

Similarly, it is also the case that an entity could not bring itself forth other than through self-construction and self-delimitation. A thing could be present by other means—for example, if it were constructed or delimited by other beings, but it could not place *itself* into appearance other than through *self*-construction and *self*-delimitation. Conversely, in constructing and delimiting itself, the being necessarily brings itself forth, thus becoming open to any being exposed to the clearing, for, in constructing and delimiting itself, the entity brings itself into existence, thereby coming to "stand out" from the rest of reality. The concept of existence thus assumes here a phenomenological sense that is in keeping with its Greek etymology (from *ek-*, meaning "out," and *sister*, meaning "to stand")—namely, "standing out," where standing out is to be understood in the sense of manifest distinction from a background or environment.[48]

Could it not be objected here that, while necessary for self-bringing-forth, self-construction and self-delimitation could occur without the entity thereby being brought to presence? It is certainly true that beings may produce themselves without being perceived. But in producing themselves they necessarily open themselves up to being perceived, and it is this *self-opening up to being perceived* that is signified by the concept of self-bringing-forth. It is in this sense, then, that a being that constructs and delimits itself necessarily brings itself forth. It does not necessarily present itself to a being who directly perceives it; rather, it brings itself into a space of appearance wherein it becomes open to being perceived. It is present, in the sense of being open to being perceived, though not necessarily in the sense of actually being perceived. And it is this *self-opening-up-to-being-perceived* that necessarily follows from the conjoined processes of self-construction and self-delimitation: any being that constructs and delimits itself is necessarily also open to being perceived, though this is not to say that it is at every point actually perceived.

## Physis, Technē, and Science

In the above sections, we examined the concept of physis, understood as self-production, and thus also self-bringing-into-existence. But there are ways other than physis that beings may be produced and brought into existence. In this section, we will consider two such ways, technē and science, with a view to setting forth the outlines of a general theory—of direct relevance to later discussions of biomimicry—not only of nature but also of technē and science, while at the same time looking at some important the ways that nature, technē, and science relate to one another.

The ancient Greeks distinguished between physis and technē. Physis, as we have just seen, is both the self-production and the self-bringing-forth of an entity. Beings that result from technē, by contrast, are produced and brought forth by human beings. They are *produced* by human beings in the sense that it is humans who, first, *construct* them by producing and/or assembling their parts; and, second, *delimit* them from their environment such that they come to exist as finite and separate entities. And they are *brought forth* by human beings in the sense that the production process results also in their coming to appear. A computer, for example, is produced and brought forth by humans by manufacturing and assembling its various different parts, as well as through delimiting the entity from its environment through the production of a boundary, which in turn defines what is internal and external to it, and across which various different inputs and outputs may pass.

There are, however, various entities that are brought forth neither through physis nor through technē, and that, as such, would appear to fall outside the scope of the binary distinction between the two. These entities come in three types. First, there are Atoms, understood here not in the contemporary sense of beings composed of a nucleus and orbiting electrons, but rather in the etymological sense of uncuttable physical units. These units clearly do not construct themselves, for, lacking parts, it cannot be said that their parts produce other parts of themselves, and lacking any "inside," they also do not produce a limit or boundary that separates them from what thereby becomes their "outside." An example of such an entity would be an elementary particle, such as an electron, floating freely through space.

Second, there are aggregates, whether of Atoms in the sense described above, or of self-producing entities. Aggregates differ from beings that are in the way of physis in that they neither construct themselves through the production of parts by other parts nor delimit themselves through producing a limit or boundary that sets them apart from what thereby becomes their environment. An example would be the so-called chemical soup from which life first arose, for it neither constructs itself nor sets itself apart from its environment in the way that, say, living beings do.

Third, there are the non-self-producing parts of self-producing entities. If a dog is a self-producing entity, then an example of this third type of entity would be the dog's nose, which does not produce and present itself in the manner of physis, for, though it participates in the self-production of the dog, it does not delimit itself in the sense of bringing itself to stand out as a separate and finite entity in the way that the dog itself does.

But if these three categories of entities do not exist as a result of either physis or technē, then how do they exist? How, if not through physis or technē, are they brought into existence? The answer, I contend, is that they are brought into existence through science; it is through science that Atoms, aggregates, and the non-self-producing parts of self-producing entities are brought into existence. This is not to say that such entities are *constructed* through science, in the sense of having their parts produced or assembled by scientists. But they are *delimited* and *brought forth* through science, for in every case it is through science that the entity in question is cut out and thus made to "stand out" (*ek-sist*) from what thereby becomes its environment. So, whereas in the case of both physis and technē, constructing, delimiting, and bringing forth belong together, in the case of science, delimiting and bringing forth occur in the absence of construction; scientists delimit and in so doing bring forth the entities with which they are concerned, but they do not construct them in the sense of producing or assembling their parts.[49]

Could one not object here that the three types of entities brought forth through science were not brought into existence through science, but only, as it were, uncovered or discovered through it? From this perspective, noses exist whether or not we delimit them from the rest of the body, for example, in order to study their role in respiration; the chemical soup from which life first emerged was there long before we came into existence and began to take an interest in its component substances and processes; and free-floating elementary particles likewise exist independently of any scientific study we may make of them.

But if an entity must be delimited from its environment in order to exist, this objection is mistaken. Science, as its etymology tells us, does not simply look at or observe preexisting entities; it cuts them out or delimits them and only in so doing brings them into existence, where existence is understood in the etymological sense of "standing out" with respect to a background or environment. But none of the three types of entities presented above—elementary physical units, aggregates, and non-self-producing parts of self-producing beings—ek-sist in the sense of standing out prior to being delimited through the cutting action of science. Human beings may not construct these entities, and yet we do cut them out from what, through the very process of science (understood in its verbal, etymological sense), becomes other or different from them, thus making them stand out or ek-sist.[50]

The three types of entities brought forth through science do nevertheless possess a special ontological status prior to their coming into existence and which calls for the coining of a new word: "ensistence." Entities brought forth through science may not *exist* prior to the cutting action of science, but they do *ensist*, where to ensist is precisely not to stand out, and thus to appear, but rather *to be concealed within and to remain immanent to some other reality*.[51] Prior to quantum physics, for example, such entities as quarks and leptons only ensisted, in the sense of being concealed within and remaining immanent to some other reality, but after having been cut out and thus brought forth by quantum physicists, these particles can be readily encountered in the world, especially in the subsequent studies and experiments of other quantum physicists. To talk about ensistence is, however, fraught with paradox, for, while we may talk about a being as "ensisting," in conceiving it *as a being* we necessarily conceive it as "existing"; in speaking of quarks and leptons, the scientists who discovered them (overcoming of concealment) at the same time also cut them out and delimited them (overcoming of immanence), thus bringing them into ek-sistence."

Despite this paradox, it is nevertheless clear that the above entities brought forth through science differ radically from entities brought forth through physis and, as such, are not "natural" (in the sense of being self-producing). Entities that are in the way of physis do not ensist prior to existing, for they not only bring themselves into appearance but also produce themselves through the conjoined processes of self-construction and self-delimitation. Their self-bringing-forth is at the same time also a self-constructing and self-delimiting, in which case they clearly do not ensist prior to ek-sisting. In the case of the three categories of entities delimited through science, by contrast, they may not have existed prior to having being cut out through science, and yet they ensisted before science in the sense that they previously remained undelimited from, and thus also concealed within and immanent to, some other reality.

In view of the above, it is clear that we are concerned here with three different ways that beings may be brought forth. First, beings may bring themselves forth, as occurs through the processes of self-construction and self-delimitation characteristic of physis. Second, they may be constructed, delimited, and thereby also brought forth by humans, as occurs through technē. Third, they may be delimited and brought forth, but not constructed, by humans, as occurs through science.

Of deeper ontological significance, however, than this threefold distinction between physis, technē, and science, is the twofold distinction between, on the one hand, beings that bring themselves forth (physis), and, on the other hand, beings that are in one way or another (science or technē) brought forth by human beings. This in turn raises the question of the ontological status of the entities that belong to each side of this basic divide. If physis, as Heidegger at one point remarks, is "being itself," does that mean that beings brought forth through science and technē are not true beings, perhaps even that they are not "real?" Usually when we refer to things as not "real," we mean that they are imaginary or fictitious, existing in our minds but not in external reality, in the manner of, say, unicorns. But the beings brought forth through science and technē are not unreal in this sense; they are not only in our heads but present also in the world around us—even if only subsequently to having first been brought forth by us. As such, they possess a definite way of being and existing, albeit one that is very different from those that bring themselves forth.

What, then, is the way of being and existing characteristic of beings brought forth through science and technē, and how does it differ from that of beings that bring themselves forth? The various different ontological positions recognized by philosophy all identify "being" with a certain category (or, in some cases, multiple categories) of beings, arguing that any being that does not belong to that category (or those categories) is not a being at all, that it is unreal, though we may mistakenly believe it to be real. It is instructive to realize, moreover, that the basic distinction between being and not-being derives ultimately from Parmenides. It was Parmenides who first radically separated being from not-being and then sought to maintain the purity of being by rejecting its intermingling with any form of not-being. This approach was then taken over by both idealism and materialism, both of which sought to preserve the true beings—whether mental or material—from the not-being they decried in the ontology of their opponents. Idealism identifies being with minds and their contents, rejecting matter as anything other than just another form of mental content, and materialism identifies being with matter and its configurations, rejecting the notion that mind is anything other than just a configuration of matter.

In the present ontology, by contrast, the basic term is not simply "being," but "being itself." This latter expression is to be understood in two senses. First, "being itself" refers to what being itself is—namely, physis, as opposed to what it is not, such as, to take the two obvious alternatives, mind or matter. Second, and of greater immediate relevance, the expression "being itself" also has a verbal sense, according to which it refers to one possible way in which a being may be: *being itself* as opposed to *not being itself*.

This verbal sense of "being itself" is most readily understood when applied to persons, though, since a person is not a mere thing, one does not talk here of "being itself," but rather of "being oneself." What is meant by this familiar expression is being and existing because of one's self, generating one's being and existence from out of oneself, or, in a word, being "authentic" (from the Greek *autos*, meaning "self," and *hentes*, meaning "being" or "doing"). Not being oneself or being inauthentic, by contrast, is being and existing because of another, receiving one's way of being and existing from another—for example, by complying with pressure to conform exerted by one's peers.

But this distinction between being oneself or authenticity and not being oneself or inauthenticity may also, mutatis mutandis, be applied to things, and not just to persons. For a thing to be itself or to be authentic is for it to be and exist because of itself, to generate its being and existence from out of itself, to produce and bring itself forth. For a thing not to be itself or to be inauthentic, by contrast, is for it to be and exist because of another, to receive its being and existence from another, to be produced and brought forth by another.

This distinction between authenticity and inauthenticity corresponds to the distinction between, on the one hand, beings that are by way of physis, and, on the other hand, beings that are by way of science and technē. Whereas beings that produce themselves in the manner of physis owe their being and existence to themselves, beings that are brought forth by human beings, whether through science or technē, owe their being and existence not to themselves, but to other beings. But this distinction between authenticity and inauthenticity, between physis and its opposites (science and technē), does not correspond to the traditional distinction between beings that "are" and beings that "are not," and thus also to the distinction between being and not-being, reality and unreality. Beings brought forth through science and technē possess both being and existence, and, in keeping with this, one may say both what they are and that they are; and yet their being and existence comes not from themselves, but from us, for it is we who in one way or another bring them forth. They may be inauthentic, in the sense of not being themselves, but that is still a definite way of being and existing.

## PHYSIS AND THE NATURAL SCIENCES

### Physis and Natural Science

The question posed at the outset of this chapter was that of the nature of nature. The answer I have put forward is that nature is physis, where physis is understood as self-production. But self-production has thus far been only conceptualized and theorized, not applied to any actual beings. It remains to be seen, in other words, which beings—if any—actually are self-producing. In order to answer that question, however, we must first consider the relation between physis and natural science.

The concept of physis is not a scientific concept, for it can only be thought by means of philosophy, understood not as metaphysics but rather as what Heidegger calls "thinking."[52] The basic reason science cannot think the concept of physis is that it is in the very nature of science to bring beings forth by delimiting them. That beings may bring themselves forth by delimiting themselves is not a possibility to which science, as such, is open. A human being who is also a scientist may conceive of the concept of self-bringing-forth, and thus of a being that brings itself forth, but they are doing so as a philosopher or thinker, not as a scientist.[53] Science may allow us to conceive the forces or mechanisms by which Atoms combine together to give rise to composite entities, but it does not allow us to think the self-bringing into existence of beings.

Further, the concept of self-production even appears to the scientist as self-contradictory, for, assuming that everything is produced by something else, they reason that for a being to produce itself it would have to *exist before it exists*, which is a logical impossibility.[54] When a scientist thinks about a living being, for example, they typically see it as produced by its genes, by its environment, or by some combination of the two, but never by itself, and the same principle applies to all self-producing beings.[55] These considerations explain why scientists are perfectly ready to employ such concepts as "self-regulation" or "self-organization," but not "self-production." Beings that already exist, they reason, may regulate or organize themselves subsequent to having been produced by something else, but the idea that beings could *bring themselves into existence* remains unthinkable for the scientist.

To say, however, that physis is unthinkable by science should not be taken to imply that it is thinkable by philosophy, at least if philosophy is understood metaphysically in terms of the exercise of pure reason. Kant distinguishes between the *logical* use of reason, which aims to uncover the conditions of things, bearing in mind that these conditions will themselves be conditioned by other conditions, and the *pure* use of reason, which aims to uncover the *unconditioned* conditions of things.[56] So, whereas the logical use of reason may seek to uncover the cause of a specific phenomenon, the pure use of reason seeks to go back to the very beginning of the causal chain and to uncover the uncaused cause—such as God, the soul, or the beginning of the world—which underlies and completes that chain. In the case of physis, however, there is no uncaused cause responsible for the subsequent chain of causes, for the cause of a *physei*-being lies not in *other* things, whether caused or uncaused, but in itself. Physei-beings escape the linearity of both the pure and the logical uses of reason by causing, or rather producing, themselves.

Another important feature of the thinking of physis is that it overcomes the radical divide between metaphysics and physics, philosophy and science. From Plato to Kant, a divide opened up between the metaphysical world accessible only to pure reason, and the physical world accessible to the senses, and this meant that philosophy and science ultimately dealt with two quite separate worlds. One could engage in speculation about the paths that might join them together, but separate worlds they remained. When being itself is conceived as physis, by contrast, there is only one world, the world in which things come to presence (i.e., the clearing), with philosophy thinking the self-bringing to presence of things

(i.e., being itself as physis), and science bringing things to presence through artificially delimiting them. In keeping with this, when philosophy thinks physis, what
is thought is not some sort of "thing behind the thing," the Idea behind the physical reality, the noumenon behind the phenomenon, the thing-as-it-is-in-itself
behind the thing-as-it-appears-to-us, but simply *the thing being itself*—that is to say,
the process of self-production (physis) as and through which the thing exists.

Nevertheless, the fact that for the thinking of physis there is only one world
opens up the possibility that philosophy, in the sense of thinking, may be articulated with science. Indeed, physei-beings, whose being is thought by philosophy,
may be cut up or analyzed by science, and it is through this process of cutting
physei-beings up that the articulation of philosophy and science occurs. If we
understand the concept of articulation in its etymological sense as "separating
into parts" (from the Latin, *articulus*, meaning "part"), we may even say that it is
science itself that "articulates" the concept of physis thought by philosophy, for
science may separate self-producing beings into their constituent parts while, by
means of the logical use of reason, analyzing the causal chains that occur within
them.[57] But this articulation is more than just a mere possibility: if we are to identify those beings that bring themselves forth, it is also a necessity, for it is only
thanks to science that we may observe and identify the production of an entity's
parts by its other parts (self-construction), as well as the different ways in which
the entity produces a boundary or limit that sets it apart from its environment
(self-delimitation). Conversely, if scientific analysis does not allow us to identify
conjoined processes of self-construction and self-delimitation, then no self-
producing entity can be said to be present.

Science does much more, however, than simply help us identify those entities
that are self-producing. It also makes it possible to analyze the details of the process of self-production—for example, by allowing us to identify a self-producing
entity's internal systems or the materials of which these systems are composed,
all of which, prior to being revealed by science, should not strictly speaking be
described as "produced," in the sense of brought into *existence*, but rather as
"enduced," brought into ensistence. Likewise, science makes it possible to analyze
the relations between the self-producing entity and its environment, including,
if they occur, various different flows—whether of materials, energy, or
information—into and out of the entity in question. All of this in turn makes
possible comparative analyses, whether between individual beings, between different types of beings, or even between ways of self-producing characteristic of

entire regions of being, such as the physical or the biological. In this manner, it may be possible to identify different ways of being within the fundamental way of being that is self-production. With this in mind, let us now consider the relation between physis and the various natural sciences.

## Physis and Physics

As we saw earlier, the concept of self-production or autopoiesis, as it first emerged in the contemporary era in the thought of Maturana and Varela, was thought to apply only to the living. This limited application was soon challenged, however, by Morin, who contended that certain purely physical beings are also self-producing. In view of this, let us now take a critical look at Morin's analyses of the relation between physis and physics.

We have already seen that Morin interprets poiēsis as making or creating, but not as bringing into appearance. This leads to a difficulty regarding the ontological and existential status of the entities of which self-producing beings are composed. On the account I put forward above, these component entities only exist subsequent to having been delimited and brought forth through science, thereby transforming their prior ensistence into existence. But, lacking any account of science as a specific way of bringing forth, as well as the distinction between ensistence and existence, Morin offers a different account of the ontological status of these entities.

The most basic self-producing beings, Morin tells us, are atoms, understood in the contemporary nonetymological sense of beings composed of a nucleus and orbiting electrons. There is, therefore, no problem in identifying the entities of which self-producing beings at higher levels of organization (e.g., organisms) are composed; they are composed of atoms, understood as lower-level self-producing beings. But what, then, of the ontological status of the elementary subatomic particles of which atoms are composed? Morin draws here on quantum physics to claim that they possess "hardly any being" (*à peine de l'être*) and only "flashes of existence" (*des clignotements d'existence*).[58] Quite what he means by this—and quite why he says it—is not entirely clear, but the general idea seems to be that, since particles behave "chaotically" and so lack the "organization" characteristic of self-producing beings, they are somehow less real.

The basic problem with this position is that it is far from clear why the distinction between not being (or hardly being) and being, not existing (or existing only in flashes) and existing, would map onto the distinction between chaos and organization.[59] To be or exist in a state of chaos is still to be or exist, just as to be in a state of organization is to be and exist. But if the concept of chaos, which derives from the thought of the early Greeks (especially Hesiod), fails to provide an adequate ontological characterization of that which preceded the self-bringing-into-existence of the first true beings, there is, I believe, another early Greek concept that is up the task: Anaximander's *apeiron*.

According to Anaximander, it is from the apeiron that beings arise, and to the apeiron that they eventually return. But what is the apeiron? The word *apeiron* is formed through combining the privative *a-* with the word *peras*, meaning "limit." The apeiron, it follows, is not the chaotic or unorganized but the "unlimited," that in and around which no limits are present. But if to exist requires the presence of limits, for it is only after a being has been delimited in one way or another—whether by itself or by another—that it can ek-sist in the sense of "standing out" from what it is not, then it is clear that nothing actually exists prior to the emergence of beings from out of the apeiron. The reason, then, that elementary particles do not exist, at least prior to having been delimited through science, is not, as Morin seems to think, because they are not organized, but because they do not possess any "limits." Science may be able to analyze the apeiron in such a way that it is carved up into various different elementary particles, but the elementary particles thereby disclosed do not bring themselves forth, as self-producing beings do. Instead, they are brought forth through science, which, in delimiting them, takes them from a state of ensistence—concealed immanence—into one of existence.

It remains to be seen, however, which physical entities—if any—bring themselves into existence from out of the apeiron and which, by way of contrast, are brought forth by us. It would be beyond both the scope of this book and the knowledge and abilities of its author to attempt anything like an exhaustive analysis of all physical entities with a view to working out which, if any, are self-producing, so I shall thus limit myself to a brief discussion of the three types of entity on which Morin's analysis focuses: atoms, vortices, and stars.

Let us begin with atoms. Are they self-producing beings or just aggregates of matter artificially delimited and brought forth through science? The parts of an atom are clearly not produced by other parts, at least in the sense that protons, neutrons, and electrons are not generated within atoms and, in keeping with this, may all be observed outside of atoms. But, just as human technē may produce a new entity by assembling other entities that were previously just lying around, is it not possible also that atoms may "construct themselves" in the sense of assembling themselves from out of other more basic entities? To answer this question in the affirmative may at first appear to go against the concept of self-construction—the production of parts by other parts—set out above. But while the parts of an atom are not made or created by other parts, they may be said to produce other parts inasmuch as *they make them into parts*. Having previously been separate, the parts of which the atom is composed join together in ways that produce new properties in one another, such that these parts, *conceived as parts*, may be said to produce one another. An electron, for example, behaves very differently as part of an atom—that is to say, when brought into the orbit of a nucleus, from how it does when floating freely through space.[60] Atoms, it would seem, do indeed self-construct.

What, then, about self-delimitation? The various different parts of an atom do not just modify one another's behavior; by means of the strong force operative in the nucleus and the electromagnetic force between the nucleus and the electron cloud, they hold one another together. This reciprocal holding together of the parts of the atom in turn produces a limit in the sense of a distinction between that which is held together as part of the atom and that which is not. Atoms, it would seem, are also self-delimiting. And, as both self-constructing and self-delimiting, they are also self-producing.

The second key example Morin presents of physical self-production are vortices. As in the case of atoms, the parts of a vortex do not produce each other, at least in the sense that the parts of which the vortex consists (e.g., water molecules) are not produced by other parts of the vortex, but they are produced *as parts* inasmuch as they modify each other's physical properties (specifically the course of their flow) in such a way that they join up to form a dynamic circularly organized entity. Vortices, it would seem, are indeed self-constructing. Further, their circular organization in turn means that the vortex closes in on itself, thus producing a limit that separates it out from what thereby becomes its environment and that may be contrasted with a linear flow (e.g., a stream) that is open at both ends, and so not self-delimiting. This limit also manifests itself empirically from a thermodynamic perspective; on the internal side of the vortex, entropy is kept low, whereas on the external side it increases. Vortices, it appears, are not only self-constructing, but also self-delimiting, and therefore self-producing.

Morin's third key example is stars. Self-construction in stars occurs when nuclear fusion reactions produce new atomic nuclei and therewith also large quantities of free energy that in turn produce new fusion reactions. Whether in proton-proton chains, in which the product of one reaction provides the material for the next reaction in the chain, or in the CNO cycle, in which hydrogen atoms power a cycle in which a nucleus is endlessly transformed into carbon, nitrogen, and oxygen, releasing helium and free energy in the process, the parts of stars are manifestly produced by the star's other parts. Further, these parts together give rise to a sufficiently strong gravitational attraction for them to be held together, thus preventing the violent energy produced by the nuclear reactions in the core from tearing the star apart. This in turn produces a limit that separates the star out from its environment, such that it may indeed be said to delimit itself.

To see the ontological importance of this limit, it is instructive to contrast stars with various other instances of nuclear chain reactions. In the case of big-bang nucleosynthesis, by which some nuclei heavier than hydrogen were produced in the immediate aftermath of the big bang, chain reactions occurred, but the rapid expansion characteristic of the early universe led to a major reduction of its temperature and density, such that the resulting nuclei ended up going their separate ways and did not come to participate in the being and existence of higher-level self-producing entities (i.e., stars). Something similar also occurs following the detonation of nuclear bombs. A chain reaction ensues, but no limit is produced, the result being that the explosion rapidly exhausts itself and no enduring entity comes into being. In the case of nuclear reactors, by contrast, the chain reaction is contained within a limit or boundary. However, this boundary is not produced naturally, by self-delimitation, but instead by human technē; parts may produce parts, but they do so only within an artificial limit or boundary. Only in stars, then, do the nuclear reactions themselves give rise to a limit that in turn allows the reactions themselves to continue, thus giving rise to a being that produces itself.

## Physis and Biology

There can be little doubt that the concept of self-production or autopoiesis first arose in the context of biology, more specifically in the work of Maturana and Varela. But while Maturana and Varela may have been mistaken to identify self-production with life, this is clearly not to say that life is not self-producing; it is but one mode of self-production among others.

In keeping with Heidegger's idea that a renewed thinking of being would emerge from within the roots of cybernetics, Maturana and Varela's concept of autopoiesis arose in a context dominated by the application of cybernetic concepts to living beings. As an example, consider Richard Dawkins's speculations on the origin of life. Dawkins suggests that selfish replicators, the ancestors of genes, came into existence before the first single-celled organisms. Their selfishness, he tells us, consists in their "aim" of replicating themselves as widely as possible, where "aim" is but a figurative way of describing the phenomenon whereby those molecules most successful at replicating themselves are naturally selected.[61] Life, Dawkins then tells us, emerged when these primitive replicators underwent mutations that allowed them to construct "robot vehicles" around themselves, which in turn increased the replicators' likelihood of making copies of themselves and were thus favored by natural selection.[62] At this precise moment, the selfish replicators became genes, understood as entities that generate a phenotype, itself understood as a cybernetic machine whose purpose is to satisfy the "aims" of its genes.[63] Living beings, on this view, are cybernetic machines created and controlled by their genetic makers and masters.

To understand how the concept of autopoiesis makes possible a radical break with this view of the origin and nature of life, let us consider the work of Stuart Kauffman, for it not only stands in close proximity to that of Maturana and Varela, but goes further than theirs in its attempt to provide an explanation of how it was that life first came into existence. According to Kauffman, abiogenesis occurred not through the replication of macromolecules (Dawkins's selfish replicators), but rather through a process he calls "autocatalytic closure," which may be analyzed and understood in terms of the two basic traits of physis set out above: first, self-construction, which in this instance takes the form of autocatalysis (i.e., chemical reactions whose products are catalysts for reactions of the same type or for coupled reactions); second, the setting up of a limit or closure, such that the autocatalytic reactions do not grow and extend indefinitely, but instead come to form a closed network in which, as Kauffman puts it, "each molecule's formation is catalyzed by some other molecule in the organization."[64]

From this perspective, genes are not the "creators" of life, for they are but complex organic molecules that arise either through the internal differentiation of autocatalytically closed entities, or through subsequent incorporation into such entities.[65] Further, like the other components of living beings, genes participate in the process of autocatalytic closure characteristic of living beings, rather than being their preexisting masters and makers. So, whereas Dawkins considers living beings as products of other preexisting entities, the selfish replicators, Kauffman holds that, "with autocatalytic sets, there is no separation between genotype and phenotype. The system serves as its own genome."[66] The system, in other words, generates or produces itself.

Kauffman further contends that at a certain point in the growth of autocatalytic sets they spontaneously break in two and, in so doing, self-reproduce.[67] The beings that result from this process of self-reproduction may thereafter come into competition for existence, and this in turn makes possible the natural selection of those traits that favor either the self-production or the self-reproduction of the individual. In this manner, the first autocatalytically closed entities were able to evolve into the sort of highly differentiated single-celled organisms we see around us today.

It is also important to note that, as organisms evolve and thereby often gain in complexity, the twin processes of self-construction and self-delimitation may take on quite different forms to those described by Kauffman. Self-construction may involve not only the production of organic molecules but also of internal organs, circulatory systems, skeletons, and so on. Likewise, the limits or boundaries of the organism may include more than just obvious physical boundaries, like the cell wall or skin, for these boundaries may also be determined by such things as the nervous system and the immune system, both of which play a crucial role in producing and maintaining the boundary between self and other. Nevertheless, at all of these different levels of organization and in all these different ways, self-construction and self-delimitation are present, in which case all organisms may be characterized as self-producing.

It remains to be seen whether, among the many phenomena studied by biologists, there exist autopoietic beings other than organisms. Maturana suggests that there may exist what he calls "higher-order" autopoietic beings composed of multiple organisms, such as insect societies.[68] He does not, however, explore this idea in any detail. With this in mind, let us now turn our attention to the question of whether societies of organisms—including insect societies, such as societies of ants, termites, or bees, as well as other animal societies, such as societies of wolves or chimpanzees—are self-producing.

There can be little doubt that parts of these societies produce other parts. In the case of many insect societies, this production is centralized, in that a single part of the society—the queen—is involved in the production of the other parts, including the workers. Similarly, in some mammalian societies, reproduction is restricted to the alpha pair, with the other members being their offspring. In other mammalian societies, by contrast, reproduction occurs between multiple males and females. But, while it is very often the case that reproduction will only occur between existing members of a society, it is certainly not the case that societies always and only delimit themselves through the closure of their reproductive networks. In the case of African wild dogs, for example, the females leave their birth pack at age two and a half in search of another pack to join.

Self-delimitation in insect and animal societies, it would seem, must arise in another manner. And this, I suggest, is the identification of fellow members of one's society and in one way or another bonding or cooperating with them (including through reproducing with them), though this is of course not to say that bonding and cooperation is in every case the *only* mode of interaction. Identification of fellow members could involve direct recognition of individuals, as occurs in some mammalian societies, recognition on the basis of signs that differentiate members of the group from others, as happens in ants through a blend of hydrocarbons present on the cuticles, or even through the simple awareness of spatial proximity, as with schools of fish. It is important to realize that the production of structures or constructions, such as nests or dams—which of course requires cooperation—may also participate in the self-production of the society, for these abiotic structures are produced by, and productive of, its biotic components. Some biological societies, then, are composed of both biological individuals and the abiotic structures they build.

The environmental philosopher Warwick Fox has suggested that species, too, are autopoietic, though he does not provide a detailed explanation or justification of this claim.[69] So, are species autopoietic? Self-construction, understood here in terms of parts producing other parts, would indeed appear to be present in species, for species persist over time as the biological individuals of which they are composed produce other biological individuals, and so on, potentially ad infinitum. In human beings, for instance, new "parts" of the species emerge through the sexual reproduction of other "parts" of the species.

As for self-delimitation, it arises in species when, for one reason or another (e.g., geographical separation, the entering of a new niche, genetic polymorphism), members of certain populations come to reproduce only with one another, thus not only giving rise to a certain genetic and phenotypic homogeneity but also, over time, eliminating the possibility of reproduction with individuals who, from that point on, no longer belong to the same species. Thus, just as life, on Kauffman's view, first emerges when autocatalytic processes come to form a closed network of chemical reactions, species first emerge when there forms a closed network of sexual reproductions. To this it may perhaps be objected that some apparently separate species may occasionally interbreed—lions and tigers, for example, thereby giving rise to "ligers." The offspring of such encounters are usually sterile, however, and in those cases where they are not, the next generation of offspring will usually be weak or fragile in some respect and therefore highly susceptible to elimination by natural selection. When interbreeding is successful, by contrast, we may conclude that the separation was only apparent, and that they belonged to the same species all along.

But if organisms, societies, and species are all autopoietic, not all of the entities studied by biologists are autopoietic. Populations, in particular, are not. They may at first sight appear to construct themselves, for individuals within a population may reproduce through interbreeding (parts producing other parts), but they do not satisfy the second condition, for populations do not delimit themselves in the sense of establishing a limit or boundary that closes them off from other populations. On the contrary, it is the scientist who establishes the limit or boundary that delimits a population such that it may be treated as a separate entity distinct from the rest of reality. This is not to say that scientists do not have reasons to establish boundaries between populations; populations are usually differentiated on the basis of geographical location. But the geographical features that retain a population within a certain area are not produced by the population itself, and, in that case, the population cannot be said to be self-delimiting. Through interbreeding, the population may, over time, come to delimit itself in the sense of forming a closed network of sexual reproductions, but the entity that thereby produces itself would not be a population, but a species.

Organisms, societies, and species, are, I believe, the only self-producing biological entities. But rather than pursue this line of inquiry any further, let us turn our attention to the question of what, if anything, is specific to biological instances of autopoiesis. The answer, I suggest, lies in the original meaning of the ancient Greek word *logos*. This word usually refers to reason or speech, but its original meaning, as Heidegger often reminds us, is "gathering."[70] What is gathering? To gather is not simply to bring together; it is also to separate out. When one gathers wood for a fire, one does not simply bring pieces of wood together, for one will also separate out other pieces—for example, those that are too wet, too small—retaining only those suitable for the fire. To gather, *legein*, it follows, is both to *collect* (from *com-*, meaning "together," and *legere*, meaning "to gather") and to *select* (from *se-*, meaning "apart," and *legere*, again meaning "to gather").

The example of gathering wood for a fire is, however, potentially misleading. Gathering does not just involve simply selectively bringing things together into a heap or bundle, but a certain arrangement of the things gathered together. A collection of stamps, for example, may be gathered in the sense that the stamps are selectively brought together over time, but the stamps will also be arranged in a certain manner, with any new stamps being expected to fit into the existing arrangement (though they may also call for or suggest a new arrangement). There is another respect in which the example of gathering firewood may be misleading. Gathering does not have to be a process carried out by a subject aware that it is carrying out the action of gathering; gathering may simply occur *within* the being in question, but without being carried out by the being qua subject, and so quite differently from when humans gather wood for a fire.

With a view to demonstrating that legein is a general feature of biological entities, let us start by considering living beings. Now, it is not difficult to see that legein occurs within living beings, for various things, most obviously nutrients, are selected for incorporation into living beings, whereas others are not selected, are avoided, or are expelled as wastes. Further, in keeping with the analysis of legein presented above, that which is selected is not simply held together in a heap or bundle but is integrated into the characteristic organization of the being in question.

If legein occurs within living beings, it is a result of natural selection, for beings capable of gathering are more likely to endure and reproduce than those that are not. An important consequence of this is that the first autocatalytic sets studied by Kauffman are in fact only physical instances of self-production, not yet biological instances. Autocatalytic sets may have been the precursors of living beings, but, lacking legein, they were not yet alive. It was only when legein arose, through the operation of natural selection on heritable variations, that life first emerged.

The fundamental characteristic of living beings, it follows, is neither self-production nor reproduction. Both self-production and reproduction are necessary conditions for living beings to come into being, but they are not sufficient conditions. Self-production is necessary, for living beings are a type of self-producing being. And reproduction is also necessary, since it is only subsequent to reproduction that those entities in which legein is present may be naturally selected, and those in which it is not eliminated from existence. But reproduction alone does not guarantee legein, for Kauffman's autocatalytic sets could reproduce, but, at least initially, lacked legein. Conversely, legein may be present in a being that is incapable of reproduction. Sterile organisms, such as mules, are not alive on account of their self-production, for some nonliving beings also produce themselves. And they are not alive either on account of their ability to reproduce, for they cannot reproduce. What makes them alive is, rather, the presence of legein. Legein is of course only present in the mule because of a long line of successful reproductions in its phylogenetic past, but, provided legein is present in the mule, the mule is a living being, even though it is incapable of reproduction.

As for societies of organisms, legein manifestly occurs in insect societies that construct collective dwellings, such as bees, ants, and termites, for they gather together such things as nectar, leaves, and wood, which are then integrated into the characteristic structure of their dwelling. But social forms of gathering occur in other ways, too. Consider a school of fish, the members of which may gather together on arrival of a predator and then move in unison in such a way that they avoid being eaten. In this case, the members of the school are selecting certain positions within the collective body of the school while avoiding others. What is gathered together in this instance are not abiotic nutrients or materials, but the biotic members themselves—a widespread phenomenon that occurs also when a society's biotic members hunt in packs, gather together for warmth, or fly in energy efficient formations.

In the case of species, gathering or legein takes the form of natural selection itself. Via natural selection, certain traits are selected and integrated into the individuals of which the species is composed, whereas others are eliminated. As the above discussion of natural selection implies, this form of gathering operates at a deeper level than that of organisms and societies, for it is the condition of possibility of the latter; natural selection is that form of legein that makes possible all others, selecting for other forms of legein in organisms and societies. If ants can gather together materials for nests, and if wolves can gather together to hunt prey, it is because the ability to do so arose within one or more individuals and was then transmitted, via reproduction, to others, with those beings lacking the specific mode of legein in question being eliminated via natural selection.

Legein is entirely lacking, however, from beings whose self-production is purely physical. These beings may hold themselves together thanks to the setting up of a limit or boundary, but the process of legein or gathering—understood as selective bringing together—is wholly absent. Atoms acquire and shed electrons without any selection being made between them. Vortices indiscriminately suck in anything that crosses their path (e.g., people, trees, boats, houses). And stars will attract any other entity as a result of their gravity, even when doing so is "fatal" to themselves, as occurs when stars in binary systems implode on collision. The basic reason why legein is absent from purely physical self-producing beings is that they do not reproduce and so do not evolve via natural selection. Natural selection operating on heritable traits is the basic form of legein that generates all others, and, since it is absent from the purely physical realm, so, too, are all other forms of legein.

But if legein is present in organisms, societies, and species, how are we to differentiate between the respective forms of legein that occur in each of them and thus between the three types of biological entity in question? With a view to answering this question, let us note that biological entities exist in a world of constant flux and change, and that it is only by responding appropriately to any relevant changes (or "perturbations," as I shall call them) that they are able to endure—a capacity entirely absent from purely physical autopoietic beings. But the manner in which organisms, societies, and species respond to perturbations is in each case different, and it is these differing manners of responding to perturbations that make it possible to differentiate between the respective forms of legein present in each of them.

Let us start by considering organisms. Their distinctive trait is that they possess the ability to *detect and respond* to perturbations, whether external or internal, including changes in luminosity, nutrient availability, and chemical concentrations. This is true not just of animals possessing nervous systems, but of all living beings. Plants, for example, may detect predators feeding on their leaves and respond by producing unpalatable compounds, just as they may detect sunlight and respond by growing in its direction. In order to distinguish this ability to detect and respond to perturbations from mere legein—which, as we will see later, does not necessarily involve detection and response to perturbations—I will henceforth speak here of "cognition." Further, when the response to the perturbation detected involves not just a mechanical reaction, but a further process of legein by which an appropriate response is selected, I will say that the entity possesses "intelligence" (from *inter-*, meaning "between," and *legein*, meaning "gathering").

Societies also detect and respond to perturbations and, as such, may be said to cognize, sometimes intelligently. Ants are collectively capable of working out the optimum path to a new source of food, not because some clever individual has solved the problem and communicated their answer to the other members of the society, but simply through each individual ant following a simple set of rules, which, when followed by all, allow the optimal path to be selected. Biologists talk in these instances of "swarm intelligence," for it is not the individual, but the social swarm, that is detecting and responding intelligently to perturbations, such as the emergence of a new food source.[71] Just as individual neurons firing according to a simple set of rules in the brains of mammals may collectively give rise to intelligent responses at the level of the organism, so individual organisms following simple sets of rules may likewise give rise to intelligent responses at the level of societies.

Other ways in which societies detect and respond to perturbations make greater demands on the cognitive powers of the individuals involved. The direction of travel of an individual within a social group may be determined both by their own direct perception of where best to go and by imitating the movements of their immediate neighbors. Through the limiting influence of the latter on the former, the "many wrongs" of individual intelligence are suppressed, and the group is on average capable of determining a better path than any particular individual.[72] But, despite the greater role of individual intelligence here than in the cases of swarm intelligence described above, it is nevertheless the group, and not the individual, that selects the path that it is to take; the selection process is a holistic phenomenon that only emerges when the individuals of which the group is composed act together in the manner described above.

But if biological societies resemble individual organisms in that they also cognize, what is the difference between the two? The answer is that the ways that societies cognize depend entirely on the cognitive apparatus of the individuals of which they are composed. It is individual ants, for example, that detect the presence of food and respond by laying down chemical traces for other ants to follow—a form of behavior, that, when generalized, allows for the optimum path to be selected. The society itself, however, has no cognitive apparatus specific to itself, no sensory organs, nervous system, or brain. So, while in cases of social intelligence it is the society, not the individual, that detects and responds to perturbations and in so doing selects the appropriate course of action, it does so only by means of the cognitive apparatus present in the individuals of which it is composed. In the case of organisms, however, the ability to detect and respond to perturbations is present at the level of the organism itself.

Let us now turn our attention to species. Organisms and societies, as we have already seen, possess the ability to detect and respond to perturbations, and thus also cognition; this ability is a product of natural selection operating on heritable variations, usually genetic. In the case of species, however, the ability to respond to perturbations is not a product of natural selection. On the contrary, it is natural selection itself that allows species to respond to perturbations. Natural selection, from this perspective, operates within species in such a way as to eliminate those traits that reduce the fitness of the individuals—and in some cases also the societies—of which the species is composed and to increase the presence of traits that increase their fitness. This in turn allows the species as a whole to endure, for if traits that reduce the fitness of the individuals and societies of which it is composed were to proliferate, the species as a whole would be endangered. If, for instance, an extraordinary set of coincidences meant that strongly detrimental genes were to spread throughout the species—say, because lightning just so happened to strike down all those members of the species that did not possess them—then the enduring existence of the species would be endangered.

Natural selection does not allow species to respond to all perturbations, however. Some external perturbations (e.g., habitat destruction, the emergence of a new disease) may be so great that the species is unable to respond, in which case it will go extinct. Even internal perturbations may put the species in danger. A new gene that made a predator species so successful that it drove its sole species of prey extinct would itself face extinction. But something similar is true also of organisms and societies; there are perturbations that surpass their capacity to respond, in which case they perish. But that does not mean that responsiveness in organisms and societies does not, generally speaking, allow its bearers to endure, and the same goes for natural selection within species.

Further, there is a clear analogy between the ways that organisms and species respond to perturbations. Organisms possess a repertoire of possible behavioral responses that cover a range of different perturbations. Similarly, in the form of genetic and corresponding phenotypic diversity, species also possess a repertoire of possible responses that cover a range of different perturbations. Just as an individual organism may respond to food being available at a greater height than before by stretching out its neck, a species may respond to food being available at a greater height than before through the natural selection of longer necks, as occurred in the case of giraffes.

There is a major difference, however, between organisms and societies on the one hand, and species on the other. Unlike organisms and societies, species cannot *detect* perturbations. Whereas an individual giraffe may respond to food being located higher than before by first detecting the food at a higher level and then responding accordingly, in the case of species no detection process is involved. There is, in the case of species, no perception of the need for longer necks, only a process of natural selection whereby genetic mutations that give rise to longer necks proliferate through successful reproduction, with genes for shorter necks eventually being eliminated. So, while the species may *respond* to the environmental change through the evolution of longer necks, this is not to say that that environmental change was first detected, as occurs in the case of organisms and societies. Species, we may conclude, do not cognize and so cannot possess intelligence either. Legein may occur within them in the form of natural selection, but natural selection is a "blind" process, unable, as such, either to *detect* perturbations or to respond to them *intelligently*.

## Physis and Ecology

The concept of autopoiesis is, at present, largely absent from ecology. But this does not in itself mean that there are no specifically ecological self-producing entities. With this in mind, let us consider the question of what self-producing ecological entities, if any, there are.

The self-producing biological entities discussed above fall into the following categories: they may be living beings; they may be composed solely of living beings, as is the case regarding species and some societies; or they may be composed of living beings and the structures they build, as is the case regarding other societ-ies. Ecological entities differ in that they are composed of individuals of multiple species in interaction, as well as various abiotic elements that constitute the phys-ical environment of the biotic elements. These assemblages of biotic and abiotic elements are commonly known as "ecosystems."

But are ecosystems self-producing? With a view to answering this question, it is worth noting that, from the mid-twentieth century onward, cybernetics pro-vided ecology with its basic theoretical framework. The key concept of cybernetic ecology was self-regulation. Ecosystems, it was claimed, are capable of regulating such variables as species composition, nutrient availability and flow, and temper-ature in such a way that they remain stable. From an ontological perspective, however, what was overlooked in cybernetic ecology was the question of what ecosystems are and how they come to exist in the first place; they may regulate themselves, but a thing must first exist as the thing that it is if it is to regulate itself. And, here as elsewhere, the emergence of the concept of self-production suggests a possible response to the question of the being and existence of ecosys-tems: Perhaps ecosystems produce themselves in the sense of bringing themselves into existence by constructing and delimiting themselves?

The principal objection to the idea that ecosystems are self-producing concerns the notion of self-delimitation. Indeed, there are two obvious problems to over-come if ecosystems are to be considered self-delimiting: the problem of *contiguity*, and the problem of *continuity*. The problem of contiguity is that, since ecosystems are contiguous with one another—a forest may be contiguous with a lake, the lake contiguous with a prairie—there may not be any physical gap between them. Of course, on the view of being itself as physis, an entity is not defined according to an ontology of material extension, in which a physical entity is continuously extended over a finite region of physical space, while at the same time separated from other entities by an intervening gap. Continuous material extension, combined with spatial separation, is not the criterion for ontical distinctness; ontical distinctness, which in the case of nature arises through self-production, does often coincide with spatial separation, as is the case in stars, but this is not always the case. Living beings are often physically contiguous for the duration of their existence, as with moss growing on trees or mycelium growing contiguously to tree roots.

Further, according to Maturana, multicellular organisms are higher-order auto-poietic entities whose basic constituents, the individual cells, are themselves autopoietic. Even when surrounded on all sides by other cells, an individual cell,

Maturana maintains, remains self-producing.[73] In itself, then, the problem of contiguity is not fatal. After all, the very concept of contiguity (from the Latin *con-*, meaning "with," and *tangere*, meaning "touch") implies the existence of distinct entities in contact, rather than the absence of distinct entities in the first place. It is thus logically possible that ecosystems could be distinct entities even when physically contiguous.

The analogy with a multicellular organism does, however, reveal a significant complication with such a view: whereas the cells of multicellular organisms possess cell membranes (and in some cases also cell walls), ecosystems do not produce any obvious boundary that separates them from contiguous ecosystems. A geophysical boundary, such as the edge of a lake, may provide the occasion for scientists to delimit the ecosystem, but this boundary would not appear to be produced by the ecosystem itself.

This brings us to the second problem, which I claim is fatal: the problem of *continuity*. To gain insight into it, consider the phenomenon of nutrient cycling. Now, if it is the case that ecosystems retain nutrients in such a way that these nutrients tend to circulate within the ecosystem in question, this could be thought to imply that ecosystems delimit themselves, just as the circulation of matter in a vortex delimits the vortex from its environment. In the case of the temperate broadleaf forests of Western Europe, for example, nutrients are often retained within the soil instead of simply washing away, for they are held in place by tree roots and may therefore act as a nutritional reservoir upon which the biotic elements of the system may draw. In the tropical rainforests of South America, by contrast, frequent heavy rains prevent significant nutrient retention in the soil, and the nutrients are instead retained inside complex trophic cycles. In both cases, but in quite different ways, mechanisms would appear to exist that retain nutrients within the system, such that nutrient loss to surrounding rivers, lakes, wetlands, and seas is limited to a level that allows the ecosystems in question to endure. Ecosystems, on this view, could be said to delimit themselves in the sense that in the course of their self-production they produce mechanisms that retain at least some nutrients within the system so that, when one observes them from the outside one will see some nutrients tending to circulate internally.

There is, however, an obvious objection to this argument: nutrient cycles are plural, complex, overlapping, and diverse, making it highly doubtful that the identification of mechanisms of nutrient retention allows one to affirm the existence of ontically separate, self-producing ecosystems. It may be possible to identify mechanisms thanks to which one type of nutrient tends to circulate within what, with respect to that nutrient, may be considered a single ecosystem. Another type of nutrient, by contrast, may flow freely between that ecosystem and other neighboring or even distant ecosystems, such that, with respect to this second type of nutrient, these systems would appear to be but artificial delimitations within a single, larger ecosystem. The overall picture, it would seem, is not of ontically distinct self-producing ecosystems interacting with their respective environments, but a complex web of nutrient flows, crisscrossing the habitable region of the earth.

With this picture in mind, consider also the atmosphere. Given that the atmosphere is manifestly continuous, the only possible way to divide the earth up into separate ecosystems is to exclude it. The atmosphere must be considered to belong to an ecosystem's environment in order for that ecosystem to be considered a discrete entity. When observing a forest ecosystem, for example, the gases present around the trees cannot be considered part of the system; if they were, either the entire atmosphere would have to be included within the system or an arbitrary cut-off point would have to be posited beyond which the atmospheric gases are no longer part of the system.

And yet if an ecosystem is a system of interacting biotic and abiotic components, it follows that some abiotic components must be present in the system. The mineral nutrients in the soil and the water flowing through the system are thus typically included as parts of the system. When Aldo Leopold talks of the "land," for example, he defines it as a community whose components include "soils, waters, plants, and animals."[74] But on what grounds can one exclude the atmosphere, but include the water and the soils? Why are the water and minerals flowing between the soil and the tree roots part of the system but not the gases flowing between the air and the tree leaves? A similar problem applies to marine ecosystems. There can be little doubt that water flows relatively continuously between marine ecosystems, but to exclude water from a marine ecosystem is clearly absurd. Divisions between different ecosystems would thus appear to be based on scientific expediency, not on the self-delimitation of the ecosystems themselves.

Similar problems emerge if one tries to see the limit or boundary in terms of a bounded network of interacting members of certain species. The idea that interactions between species tend to form bounded networks that, as such, delimit themselves, is highly dubious. Plant species, as Henry Gleason has observed, are often present along more or less continuous gradients, rather than being grouped together into distinct assemblages.[75] As for animals, they may move relatively freely from one part of the earth to another, even migrating from one region to another very distant region, depending on the season. This is not to deny that there are dependencies between species, and that different species—of grasses and herbivores, for example—are often found together in certain regions of the earth, but there is no evidence that these interactions produce any sort of boundary or limit that sets the ecosystem to which they belong apart from the next ecosystem.[76] Scientists do have good reasons to delimit ecosystems; as in the case of populations, the basis for establishing the division may be a geographical distinction, such as the banks and mouth of a river, which, for the scientist, marks a convenient point where the riverine ecosystem ends. But, as in the case of populations, that distinction is not produced by the ecosystem itself, but by the scientist.

The fact that terrestrial ecosystems, such as forests, lakes, rivers, and prairies, are not self-producing does not mean, however, that the concept of self-production does not apply to ecology. Indeed, as we will see, this concept applies to the earth as a whole, or Gaia, the component ecosystems of which are but artificially delimited parts. Like organisms and ecosystems, however, Gaia was not initially conceived autopoietically, but cybernetically, as a self-regulating system.[77] As a consequence of this cybernetic orientation, the principal scientific questions asked about Gaia have focused on such issues as the existence, nature, origin, supporting mechanisms, and modality (contingent or necessary) of the emergent property of self-regulation. When Gaia is viewed as autopoietic or self-producing, however, the focus turns away from the property of self-regulation and toward the foundational ontological processes of self-construction and self-delimitation by which Gaia comes into and remains in existence in the first place.

Gaian self-construction is aptly summarized by Benyus's dictum "Life creates the conditions conducive to life."[78] This is true in the sense that the biotic parts of Gaia create many of the abiotic parts as by-products of their own self-production, and these abiotic parts are in turn conducive to the existence of the biotic parts. The atmosphere of the earth, which is completely different from that of other planets, was created by the earth's living inhabitants and is in turn conducive to the enduring existence of these inhabitants.[79] But Benyus's dictum is also true in the sense that some biotic parts are directly conducive to the existence of other biotic parts. Grass is conducive to the existence of herbivores not only in the indirect sense that it contributes to the production of the oxygen that herbivores require to breathe but also in that it may be directly consumed by herbivores.[80]

For Gaia to be self-producing, however, she must also produce a boundary or limit that sets her apart from what thereby becomes her environment. But, before looking at how this occurs, let us first note that the two problems discussed above in the context of sub-Gaian ecosystems—the problems of contiguity and continuity—clearly do not affect Gaia. Unlike in the case of sub-Gaian ecosystems, there is no other entity or group of entities with which Gaia is spatially contiguous, or with which her networks of production are sufficiently entangled, as to raise doubts about her existence as a discrete entity. The sun, for example, is neither contiguous nor continuous with the earth, though light does of course travel from the former to the latter. But if Gaia is neither contiguous nor continuous with any other entity, this is not in itself to say that she delimits herself. After all, free-floating subatomic particles are neither contiguous nor continuous with one another, and yet that is not to say that they delimit themselves.

So, how, if at all, does Gaia self-delimit? The most important phenomena responsible for the self-delimitation of Gaia occur in the atmosphere, for it is the atmosphere that mediates Gaia's relations to the external environment: the ozone layer filters out harmful UV light, while also letting in wavelengths necessary for photosynthesis; low levels of atmospheric methane and carbon dioxide mean that enough heat is trapped to stop the earth freezing over, while at the same time letting excess heat escape, thus preventing the land from turning into a barren desert and the oceans from evaporating; and the atmosphere also provides protection from colliding meteorites. The atmosphere, we may conclude, is not just a product of Gaia's component parts, especially in the form of gases emitted by living beings, but a producer of those components, not only through the provision of nutrients (especially oxygen, carbon dioxide, nitrogen, and water) but also through establishing the boundaries within which other constitutive processes and entities may occur and endure.

Having established that Gaia is indeed autopoietic, let us consider the specificity of Gaian autopoiesis. Lovelock views Gaia in essentially ecological terms as a home (*oikos*) for living beings. In keeping with Benyus's dictum "Life creates the conditions conducive to life," he thinks that habitability is a product of living beings. Lovelock does not just claim, however, that Gaia does in fact provide a home for living beings; he also asserts that this is her inner goal, or telos. The teleological nature of Gaia has been called into question by other scientists, who claim that the maintenance of habitability may be but a fortuitous coincidence. Indeed, there may even have been periods in the history of the earth—so-called snowball earth periods—when habitability dramatically decreased as the earth more or less entirely froze over.[81] From this perspective, there is nothing in the interactions between living beings and their abiotic environment that either necessarily or generally leads toward the maintenance of habitability; if that is what is presently the case on earth, it is a mere coincidence.

While I see no reason to ascribe to Gaia the maintenance of habitability as some sort of inner telos, at least if by a telos one means some sort of goal or purpose Gaia is trying to achieve, I disagree with Lovelock's critics that maintenance of habitability is a mere coincidence. Habitability is maintained in a comparable way to how organisms, species, and societies maintain themselves: through gathering or legein. In the above discussion of the atmosphere, for example, it was apparent that the atmosphere does not form some sort of limit that simply circumscribes what takes place within it, for it plays a selective role as well, letting only some things in and others out. Likewise, the component parts of Gaia, including the energy received from the sun, are not just piled or bundled together but organized in such a way (e.g., into nutrient cycles) that habitability is maintained.

Moreover, like the biological entities discussed above, the presence of legein within Gaia means that she is capable of responding to perturbations, both internal (e.g., the emergence of new species) and external (e.g., changes in solar luminosity). But how exactly does Gaian responsiveness work? In living beings, responsiveness involves detecting and responding to perturbations, a capacity that arises through the action of natural selection on heritable variations. But, since Gaia does not reproduce, there is no way that natural selection can act on heritable variations, and so no way that the ability to detect and respond to perturbations, no way that cognition and intelligence, could have emerged. In keeping with this, Gaia does not possess any sensory organs capable of detecting perturbations.

But the fact that an entity cannot *detect* perturbations does not mean that it cannot *respond* to perturbations. Species, as we have already seen, cannot detect perturbations, but, by means of natural selection acting on heritable variations, they may respond to perturbations in such a way that they endure. Something similar, I claim, is true of Gaia. Natural selection is operative not only within species, and in such a way that they may endure, but also within Gaia, and in such a way that she may endure, and, since her way of being is to provide and maintain habitability, this allows habitability to endure. There is, in other words, selective pressure in favor of beings that contribute to Gaian habitability, and against beings that reduce it.

With a view to demonstrating this claim, consider Barry Commoner's observation that the first living beings would have been on a "self-destructive course," for they would have been using up any available nutrients present in their environment, while creating ever more waste products.[82] If unchecked, this course would clearly have led them to run out of nutrients, to drown in their own wastes, or to some combination of the two. But, in creating their wastes, they also created conditions conducive to the emergence of new life-forms capable of feeding off these wastes. Indeed, any beings that emerged that were capable of this would have been strongly favored by natural selection. Likewise, were this second type of living being—or perhaps a third type feeding off the wastes of the second type—capable of producing the nutrients required by the first type, they would also have been favored by natural selection, for only by "closing the circle" in this manner would it be possible for nutrients to circulate indefinitely and the system as a whole—on which all of its biotic parts ultimately depend—to endure. Similarly, the subsequent emergence of beings that interrupted the circulation of nutrients would sooner or later prove catastrophic for the system as a whole, including for the beings responsible for the interruption, who would thereby find themselves on the same self-destructive path as the first living beings. Selective pressure, in such a context, would mount in favor of the emergence of any being that could in one way or another reestablish nutrient circulation.

If we extrapolate from this example to Gaian responsiveness in general, a picture emerges. The effect of natural selection within Gaia is such that any phenomenon—and not just the emergence of new life-forms—that perturbs Gaian autopoiesis in a way that significantly reduces habitability will lead to the emergence of a strong selective pressure in favor of the emergence of life-forms capable of in one way or another putting the ecological destruction to an end. This may occur in one of three ways: suppressing the phenomenon in question, or at least reducing it to sustainable levels; nullifying the phenomenon's ecologically destructive effects; or transforming the phenomenon into something ecological constructive.

An example of the first of these would be a species that, drowning in its own wastes, developed an adaptation that allowed it no longer to produce the waste in question, or at least to do so in quantities that did not put itself in danger. An example of the second would be a species that, in the same situation, continued producing the harmful waste, but developed an adaptation that allowed it to produce another waste that, through the chemical reaction of the two, nullified the ecologically destructive effects of the first. An example of the third would be the emergence of another species capable of feeding off the wastes of the first species, such that something that initially reduced habitability would, thanks to the emergence of something else, come ultimately to participate in the maintenance of habitability.

Living beings, from this perspective, may not *always or necessarily* create conditions conducive to life, but, if and when they do not, a strong selective pressure will emerge in favor of any adaptation capable of suppressing, nullifying, or even reversing their ecological destructiveness. It follows that it is not simply a coincidence that "life creates the conditions conducive to life," but rather a consequence of the operation of natural selection operating within the Gaian ecosystem. Natural selection favors beings whose presence increases habitability, while also exerting selective pressure against beings that decrease habitability. Through nothing but the operation of natural selection on heritable variations, Gaia is thus capable of responding to perturbations (e.g., new life-forms, abiotic changes, external perturbations) in such a way that habitability is maintained. It is also important to realize that the various different responses to perturbations are not just temporary events that disappear without trace after they occur; they are retained—at least as long as they are conducive to habitability—as constituent elements of the Gaian system. This is why Benyus says that after 3.8 billion years of evolution nature has learned what is effective, appropriate, and sustainable. Over time, natural selection eliminates what is ineffective, inappropriate, and unsustainable, while at the same time accumulating knowledge of their opposites, thus ensuring the continued provision of a habitable earth.

In view of all this, we may conclude that the specificity of Gaian autopoiesis, and thus also of the mode of legein specific to ecological—as opposed to biological— autopoiesis, lies in the fact that Gaia *provides and maintains a home for life*, where by "life" I mean not only individual organisms but also societies and species of organisms.

## PHYSIS AND HUMAN BEINGS

### *The Clearing*

In the previous sections, we looked at various different self-producing or auto-poietic beings, paying particular attention to the specificities of their respective ways of being. What we have yet to consider are human beings.

Now, it may perhaps be thought that, as living beings, humans are autopoietic in the same way as all other living beings—that is to say, they cognize, in the sense of being able to detect and respond to perturbations, and that the only distinctive thing about them are the especially high levels and complex forms that intelligence takes within them. In a similar vein, it could also be thought that, just as human beings are but another type of living being, human societies are but another type of biological society, the human species, *homo sapiens*, but another type of biological species, and human beings, human societies, and the human species, but further biological components of Gaia. The danger with this approach is, of course, that of reductionism, of assuming that because human beings are living beings that there is no *ontologically significant* difference between them and other living beings.

There have been many attempts throughout history to identify or establish the ontological specificity of human beings. René Descartes famously saw this specificity as the possession of a mind, whereas all other living beings, he thought, possess only bodies. But to see all living beings as cognizing casts doubt on this. One possible response to this doubt would be to see mind not in terms of cognition, but in terms of consciousness—or, alternatively, to distinguish between conscious and nonconscious forms of mind. But this again fails to establish human ontological specificity, for there can be little doubt that at least some animals are conscious.

It has also often been maintained that the difference between humans and all other living beings is that humans alone are rational. But, again, if reason involves understanding causal and logical relations between phenomena, there can be little doubt that some animals are rational, too.[83] These and other similar arguments have in turn led to a growing tendency to see humans as but one type of animal among others, one that of course has its specificities, as all animals do, but that is not on the other side of any ontologically significant divide. Humans differ from tigers, just as tables differs from chairs, but in neither case is the difference of ontological significance.

An original alternative to the question of the ontological specificity of human beings—and one that I will take up and develop in what follows—was put forward by Heidegger. Human beings, he tells us, differ from all other living beings in that they inhabit the "clearing." But what is the clearing? In its most vague and general characterization, the clearing is the space wherein things come to presence or appear, and wherefrom they may be absent or disappear. "The clearing," as Heidegger puts it, "is the open region for everything that becomes present and absent."[84] But if this explanation makes the clearing more than just an empty word, its precise nature remains to be examined and understood.

Rather than attempting a direct answer to the question of what the clearing is, let us instead start by means of a negative approach, looking at what the clearing is not. According to Heidegger, "Philosophy knows nothing of the clearing."[85] On the face of it, this is an astonishing claim. How could it be that the space in which things come to presence remains completely unknown to philosophy? However, once we recall that thinking underwent a bifurcation after Parmenides into materialism and idealism, Heidegger's claim no longer appears quite so astonishing.

After Parmenides, only two spaces were known: a mental space containing *ideas*, and a physical space containing *matter*. Further, while in Platonism, Neoplatonism, and much Christian philosophy, ideas were contained in some sort of suprahuman space, such as the mind of God, with the advent of modern philosophy ideas were for the most part considered to be contained in the space of the individual human mind. The result is that contemporary philosophy, with its characteristic oppositions between materialism and idealism, naturalism and constructivism, realism and antirealism, knows of only two spaces.

The first is a mental space internal to the subject. In it, one finds subjective mental phenomena, including perceptions, representations, beliefs, fantasies, impressions, sensations, desires, and emotions. The main words used to describe this space are "mind" and "consciousness." The second is a physical space external to the subject. It contains material entities that exist independently of human perception and understanding, not as perceived in and by the human mind. Much contemporary philosophy consists of a debate about the relation between these two spaces. Can one space be eliminated altogether—for example, by being shown to be an illusion?[86] Can both spaces exist simultaneously, and, if so, can they in some way or other connect to each other?[87] And can one space encroach upon the other, such that the boundary between the two lies elsewhere than previously thought?[88]

What, in all these debates, has either gone unnoticed or remained an intractable problem is that *in neither of these spaces do things come to presence*. In the internal, mental space of the subject, one finds only subjective mental phenomena (e.g., perceptions, representations, sensations), not things. And in the external, physical space of subject-independent reality, there are things of a sort, but they are "things in themselves" and, as such, remain concealed, closed off from human perception and understanding.[89] Many paths have of course been proposed between these two spaces, but paths are routes for getting from one space to another; unlike the clearing, they are not spaces in their own right. It follows that whatever response one offers to the questions of the being and existence of these two spaces, and of any relations that may obtain between them, the clearing remains veiled, for neither in the internal space of subjective consciousness, nor in the external space of objective material reality, do things come to presence.[90]

If it was only after Parmenides that thinking bifurcated into idealism and materialism, such that the only spaces that could be known were mental and physical ones, it may be thought that, just as the task of thinking "being itself" anew requires us to go back to before Parmenides and make explicit the early Greek concept of physis, so an analogous approach may be taken with respect to the clearing. This is true inasmuch as the clearing is the space into which physei-beings bring themselves forth, in which case the two concepts—physis and the clearing—belong together. If mind is the space that contains ideas, and physical space the space that contains matter, the clearing is the space that contains beings brought forth through physis, though, also, it is important to add, through science and technē, conceived as alternative ways of bringing things into the clearing.

But this is not to say that we may simply go back to the early Greeks and examine what they themselves said about the clearing, for, as Heidegger was well aware, they themselves did not say anything about the clearing. They may have thought about physis, as well as, more generally, about poiēsis, understood as the bringing of things to presence, but the space into which things are brought to presence remained concealed to the ancient Greeks. In keeping with this, there is no ancient Greek word—not *apeiron*, not *chaos*, not *cosmos*, not *khóra*—that corresponds to Heidegger's concept of the clearing. There can be no question, then, of making explicit a concept that was already present in the thought of the early Greeks, but only of introducing a concept that was absent from their thought.

So, how, then, are we to go about thinking about the clearing? The obvious place to start is the thought of Heidegger, for the clearing is after all a concept that Heidegger himself introduced. This is not to suggest, however, that the most appropriate approach here is simply to offer an interpretation of Heidegger's understanding of the clearing or a reconstruction of his reflections on this concept; the present context is constituted less by an attempt to understand or reconstruct Heidegger's own thought than by an attempt to think about a space appropriate to the various modes of bringing forth—physis, technē, and science—previously analyzed. In view of this, I will in what follows draw extensively on Heidegger, but with a view not so much to exegesis as to developing his thinking in such a way that it may be coherently articulated with the thinking of physis, technē, and science set out above.

In *Basic Questions of Philosophy*, a little-known lecture course given at the University of Freiburg in 1937-1938, Heidegger presents what is perhaps his most clear and insightful analysis of the clearing. The clearing, he tells us, is a "four-fold openness . . . of the things, of the region between things and man, of man himself, and of man to fellow man."[91] Whereas materialism and idealism conceive of two rival spaces radically cut off from one another, the first of which contains things, but is closed to the human mind, and the second of which contains ideas, but is closed to things, the clearing is a single and unitary space, wherein, thanks to their respective modes of openness, encounters between things and humans may occur.

It is important to recognize, however, that the sense in which things and humans are open is quite different. Humans are open to things in the sense that they are capable of understanding their being and existence. Things, by contrast, are open in the sense that their being and existence is open to being understood by humans. Or, to put it another way, whereas humans are exposed *to* the clearing, things are only exposed *in* the clearing, for while they appear within it, they themselves have no access to it, in the sense that they are not open to the things, themselves included, that appear therein. The clearing, we may infer, is an open region into which things are in one way or another brought forth, and in which they may be understood by humans. Humans, from this perspective, possess a trait that Heidegger claims is unique: the preontological understanding of being, the ability to understand what something "is," or, as he also puts it, the ability to understand something "as" something—for example, the ability to understand the thing in front of me "as" a table or "as" made out of wood.[92]

But why, one might wonder, does Heidegger think that the openness "of man to fellow man" is also an essential component of the clearing? In *Being and Time*, he analyzes what he calls "being-in-the-world" in a similar manner to how he presents the fourfold openness of the clearing in *Basic Questions of Philosophy*. Being-in-the-world, he tells us, is a unitary phenomenon that may nevertheless be analyzed in terms of its essential components, one of which he calls "being-with,"[93] by which he means being with fellow humans.[94] In *Being and Time*, being-with is to a large extent presented in what one might call an "existentialist" sense: as a fundamental component of human existence (an "existentialia," in Heidegger's terminology). It is, Heidegger tells us, an intrinsic part of human existence to be with others, to be open to them. Even when we are alone, "being-with" is still essential to our being, in the sense that we remain constitutively open to others.[95]

But human openness to others also has an "ontological" aspect inasmuch as it is only because we are open to others that we may understand things and thus also understand the being of things. For the most part, Heidegger tells us, we understand things as "one" (*das man*) understands them. When I understand that the thing in front of me is a table, for example, I understand it in a way that is no different from how the next person understands it, and the same goes for almost anything that human beings encounter in the world; we understand things as "one" understands them. And this understanding-as-one-understands is acquired through being with others. Heidegger does not dwell on the emergence or genesis of being-in-the-world, and so, at least as far as individuals are concerned, on the question of infancy or childhood, but were he to do so he would presumably accept that, if we understand things as one understands them, that is quite simply because it was other people (e.g., parents, teachers, siblings, friends) who taught us "what's what," including, for instance, what tables are, what wood is, and what it is to be made of something else.

It follows from this way of thinking that it is not the case that human beings understand what things are on their own and then confront that understanding with the understanding of other people. Rather, the understanding humans have of things is acquired from others, with any "original" understanding of what things are being but a modification of how one understands them.[96] Openness to one's fellow human beings, then, is not only an essential component of human existence but also a condition of possibility of the understanding of things "as" things and thus of human openness to things. We may potentially understand things differently from others, but that different understanding is but a modification of how one understands things, not something we initially work out solipsistically, as it were, and only thereafter discover to be different from the understanding of others. This is why "being-with," or the openness of "man to fellow man," is essential to the clearing, understood as the space where things come to presence: only through "being-with" do we come to understand the being of things.

In view of the above presentation of the clearing, let us now turn our attention first, to the question of human specificity, and second, to the ontological significance of that specificity. Living beings, as we have seen, possess the ability to detect and respond to perturbations, or cognition, as I call it. But to be open to *perturbations* is very different from being open to *things* present in the world around oneself, regardless of how intelligently one responds to those perturbations. Further, since nonhuman organisms are closed to things, it also follows that they have no understanding of the being of things. Different perturbations may elicit different responses, but this is not the same as understanding different things in different ways and acting accordingly.

It could perhaps be objected that the above analysis shows only a distinction between, on the one hand, human beings, and on the other, living beings *in their simplest and most general nature.* There may yet be other "higher" animals (e.g., chimpanzees, dolphins, dogs) who inhabit the clearing in the sense of being open both to things and to other inhabitants of the clearing from whom they in the first instance learn what's what. Whether or not this is the case cannot be settled by philosophy alone. Just as philosophy may put forward the concept of physis, but requires articulation with science to help determine to what actual beings the concept applies, so the same is true of the clearing. Philosophy may provide the concept of the clearing, but anthropology is required if we are to show that it is a space inhabited by humans, and zoology if we are to show that it is not a space inhabited by animals.

So, what anthropological evidence is there that humans inhabit the clearing, and what zoological evidence is there that animals do not? With a view to answering these questions, let us briefly consider the work of the comparative anthropologist and primatologist, Michael Tomasello. According to Tomasello, the distinctive feature of human beings is what he calls "joint attention." This occurs when one party attends to a thing, while at the same time being aware that another party is attending to that thing. The primary indicator of the existence of joint attention in young children, Tomasello claims, is pointing, for the function of pointing is precisely to draw the attention of another party to some thing or other. And since he thinks it is humans alone who jointly attend to things, this explains why it is humans alone who engage in pointing.[97]

There is, however, a complication with this claim. Great apes may not point spontaneously, but it is possible to train them to do it.[98] Yet there is a key difference between the way great apes and humans point: the pointing of great apes is only ever imperative; they point because they want another party, only ever a human, to retrieve something for them.[99] Humans may, of course, also point because they want another party to retrieve something for them. But, as Tomasello points out, human pointing is also frequently "declarative." Humans, and young children especially, point simply to show that there is something there that is worthy of interest or to draw attention to some feature or other of the thing.[100] Further, while Tomasello does not emphasize this, human pointing may also be "inquisitive." Young children sometimes point as if to say simply, "What is this thing?" Similarly, humans may also show things to others, in the sense of presenting a thing before someone else, either in a declarative way, as if to say, "Look at this thing," or "Look at this interesting feature of the thing," or in an inquisitive way, as if to say, "What is this thing?"

The ontological significance of these specifically human forms of pointing and showing is that, through them, *the thing in question becomes open with respect to its being.* When a child points at something, this provides an opportunity for an adult to show or tell them what it is (or perhaps also for the child to show the adult they know what it is), or at least about some aspect or other of its being (e.g., what color it is, where it needs to go, to whom it belongs). Conversely, when an adult points something out to a young child, this opens up the being of the thing such that it may in one way or another be understood by the child.

By contrast, when great apes point, the thing does not become open as to its being. An ape does not point to a banana in such a way that it may learn from another what it is, show another what it is, or make a joint decision regarding some aspect of its being. For an ape, a banana is not a "thing" at all—in other words, a being existing in a space external to it and whose being it understands, at least initially, as one understands it, but simply a perturbation that elicits in it a pointing response, for the great ape has learned from previous experience that pointing in its direction will allow it to satisfy a certain desire.

It's not that animals do not learn from one another, but what they learn is not what things are, but how to respond appropriately to perturbations. When a kitten is perturbed by the desire to urinate, its mother may teach it an appropriate response, which, from the point of view of an external human observer, consists in going to urinate in a litter tray. But this does not involve teaching the kitten what anything is, such as what litter trays are, what urinating is, or why litter trays are appropriate places to urinate. Only humans engage in teaching one another what things are, for they alone are exposed to the clearing.

To say that animals are not open to things is not, however, to say that a capacity for reasoning—understood here as a capacity to understand causal or logical connections—is necessarily lacking from animal cognition. But there is a significant difference between these two types of reasoning. In the case of humans, reason operates *on things*. And in order to operate on things, the clearing is required, for without it there would not be any things. In animals, by contrast, reason operates only with respect to detected perturbations and possible responses. Animals may certainly reason in the sense of being able to establish causal or logical connections between the different elements of their experience, as occurs when intelligent birds work out the various tasks they must perform in order to obtain food. The ability to reason may well be acquired through experience and learning, rather than hardwired by genetics, but the elements of animal experience and learning are not *things*.

These elements of animal experience may, taken together, be of such great diversity and richness that, at least in some respects, human experience appears impoverished in comparison. Likewise, they may also involve feats of intelligence, whether individual or social, that surpass what most—perhaps even any—humans can achieve. But animals do not and never have inhabited the clearing; they do not understand things as things. At no point have the elements of animal experience ever become "open" and thus exposed to social categorization as this or that, making them no longer simply elements of experience, but things understood in their being. No matter how rich and diverse the experiences of animals might be, or how powerful and varied their intelligence, animals have no openness to and no understanding of things *as things*. Or, to put it in Heidegger's terms, not only do animals have no *ontological* understanding of being (i.e., an explicit *theory* of being and existence, such as materialism or idealism); they also have no *preontological* understanding of being, in the sense that they do not understand what and that things are.

What, then, may we say about the ontological significance of the clearing? It is not difficult to see that the clearing is not some new type of substance. In particular, it is not some sort of "mental" substance (e.g., thinking, consciousness) in addition to the purely "material" substances that, in its absence, would be the only ones. But while there is clearly no substance dualism here, there is an ontological duality of sorts, for there is a difference between humans and all other beings that is of great ontological significance: only to humans can things become open and thus understood as to their being. Without humans, no-thing would be "there," for the only space where things can appear, the only "there" where things can be, is the clearing, and without humans, there would be no clearing. The clearing, it follows, is not a thing that is there in the world in addition to other things, but rather the one and only space where anything at all can be and exist. Indeed, it could perhaps be said that *the clearing is the world*, at least if by "world" is meant precisely that space wherein things come to presence.

## Enlightened Naturalism

The two main philosophical concepts introduced thus far are physis and the clearing. In this final section of the chapter, I show how their articulation may give rise to a general philosophical position I call "enlightened naturalism."

The concept of enlightened naturalism may be broken down into two components, represented respectively by the noun "naturalism" and the adjective "enlightened." Starting with the former, we have already seen that naturalism in philosophy is an ontological doctrine that equates being with nature. This is true also of enlightened naturalism, but with two provisos: first, what is identified with nature in this context is not being tout court, but rather being *itself*; second, nature is to be understood as physis, and physis as self-production and self-bringing forth—or, in a word, autopoiesis.

There is also a distinctive methodological aspect to the naturalism of enlightened naturalism. While scientific naturalism holds that it is science that makes possible the study of nature, with philosophy contributing to that study only to the extent that it helps clarify, criticize, and develop scientific concepts and theories, enlightened naturalism in some respects reverses the relationship between philosophy and science. In this instance it is philosophy that first thinks nature in the sense of physis, with science contributing to the understanding of nature inasmuch as it helps us identify those beings to which the concept of physis applies, the mechanisms involved in the self-production of different types of being, how these mechanisms differ from one another, and what interactions take place between self-producing beings and their environment.[101] So, whereas in scientific naturalism philosophy contributes to the scientific understanding of nature in its more abstract and conceptual aspects, in enlightened naturalism science contributes to the philosophical understanding of nature in its more concrete and empirical aspects.

The second component of the concept of enlightened naturalism is the word "enlightened." As used in the present context, this word does of course tell us that what we are concerned with is the understanding of nature as physis, in the sense of self-production and self-bringing-forth. But the choice of the word "enlightened" is not arbitrary; it stands in close relation to the concept of physis, which does not mean simply *self-bringing-to-presence*, but also, because the clearing is the space wherein things are brought to presence, *self-bringing-into-the-clearing*. Nature, from this perspective, is, as it were, "lit up"—in the general sense of becoming manifest or appearing—in the clearing. Enlightened naturalism, then, is a theory of nature as something that by its very nature "manifests itself" or "brings itself to light" and, in that precise sense, may be described as "enlightened."[102]

There is also a methodological aspect of enlightened naturalism that corresponds to the word "enlightened." Just as I claimed above that it is philosophy that tells us what nature is, with science contributing to the application of the concept of nature to concrete empirical reality, so I here claim that it is philosophy that tells us what the clearing is, with science again contributing to the application of this concept—via the human sciences—to concrete empirical reality, as was the case in the above discussion of the work of Tomasello.

To get a better understanding of the originality and distinctiveness of enlightened naturalism, let us briefly contrast it with perhaps the two main contemporary alternatives: scientific naturalism and constructivism. According to scientific naturalism, reality consists solely of elementary particles, the forces governing their interactions, and the composite entities and emergent phenomena that arise from these interactions at all different levels and scales. Scientific naturalism also holds that the space in which all of the above occur is a physical one—in other words, a space characterized primarily by its ability to contain matter.

Constructivism holds a view that is radically opposed to scientific naturalism. It maintains that there is no such thing as "nature," for everything—or at least everything we perceive or understand—is a construct of the mind. When we look at a flower, for example, what we are in fact seeing is not a natural entity, the flower itself, but some sort of image of a flower constructed by the mind. Likewise, when we think about concepts, we are thinking about something that has been constructed by the mind, either in the sense of being built into the way the mind works (e.g., Kant's pure concepts of the understanding, like identity or causality) or in the sense of being invented by the mind on the basis of experience (e.g., the concept of a flower).

Enlightened naturalism occupies what is, at least in some respects, a middle ground between these two extremes. Scientific naturalism tells us that everything is natural, that everything is either an elementary particle or an entity or emergent phenomenon arising from the interactions of elementary particles and constructivism tells us that nothing is natural, that everything—or at least everything perceived or understood—is a construct of the human mind. Enlightened naturalism, by contrast, holds that there are some things that are natural—namely, those that bring themselves forth (physis), and some things that are not natural, or at least not wholly natural, because they have in one way or another been brought forth by us, whether through constructing and delimiting them (technē) or through cutting them out from a reality to which they formerly remained immanent (science).

But this middle ground is not some sort of hybrid of the two extremes set out above, for the space in which the entities it considers are brought forth—the clearing—is entirely different from the material and mental spaces of scientific naturalism and constructivism. Neither physical nor mental, the clearing is a "third space" characterized by a kind of generalized openness—of the clearing itself, of things to humans, of humans to things, and of humans to humans—that runs counter to the radical closure and separation of mind and matter constitutive of modern dualism. Moreover, this third space is not simply an alternative to the physical and mental spaces with which Western thought is so familiar; it is the originary and unitary space that was divided up into these two opposed spaces. And, as the originary openness of the clearing closed down, so did the openness of humans to things, of things to humans, and of humans to humans, and there thus emerged so many of the problems with which Western philosophy has since had to grapple, including the problems of the external world, of consciousness, and of other minds.

In thinking the clearing, enlightened naturalism also offers a genuine alternative to postmodernism and related positions (e.g., nonmodernism, deconstruction, poststructuralism, posthumanism). Aware of the deficiencies of both constructivism and scientific naturalism, postmodernists have sought to deconstruct the traditional duality between mind and matter, humanity and nature, by removing the sharp dividing line that previously kept these binary opposites apart. One important result is that attributes previously assigned to humans come to appear in nature and attributes previously assigned to nature come to appear in humans.[103] In tearing down this partition, however, postmodernists not only leave the basic concepts of both sides largely intact but they also fail to perceive the existence of a third space, the clearing, the concealment of which has given rise to the binary oppositions they seek to overcome.[104] This in turn gives rise to a constant need—common to all theories beginning with the prefix "post-"—for postmodernists and the like to set out their own theories in contrast and in opposition to the theories they claim to go beyond, rather than to develop a genuinely new theory, as I believe is the case with enlightened naturalism.

The initial aim of this chapter was to answer the question of the nature of nature, for it is only on that basis that we may consider what it might mean to take nature as model, measure, and mentor, as proposed by biomimicry. The answer put forward was that nature is physis, where physis is understood as self-production and self-bringing forth, or, in a word, autopoiesis. But more than this has been accomplished; I have also put forward a general ontological framework, enlightened naturalism, that covers also the question of what human beings are and how they relate to nature.

This ontological framework, which centers on the articulation of two key concepts—physis and the clearing—is, I believe, theoretically significant, for, as we have just seen, it offers a genuine alternative to all the major positions in Western thought: modern dualism, the twin monisms of idealism and materialism (or constructivism and scientific naturalism), and postmodern mind-matter hybridity. But it is also of practical significance, as it provides the overarching philosophical framework in which we may come to understand what it means for humans, seen as inhabitants of the clearing, to relate to nature as model, measure, and mentor and thus to engage in quite different forms of producing, acting, and knowing than has been the case in either modernity or postmodernity.[105] It follows that, while the primary significance of the present chapter has been theoretical, it also has practical significance inasmuch as it underpins the more practice-oriented discussions of the principles of nature as model, measure, and mentor undertaken in subsequent chapters.

As regards chapter 2, the articulation of the concept of physis with the natural sciences allows us to identify and understand those beings that fall under the concept of nature and that, as such, may potentially be taken as models, measures, or mentors. These include physical entities (atoms, vortices, stars), biological entities (organisms, societies, species), as well a single ecological entity (Gaia). Further, the above analyses of physis, science, and technē provide a preliminary indication—to be worked out in much greater detail—of how nature, science, and technics are articulated in biomimetic innovation: a natural entity, one that produces itself, is first analyzed or cut up through science, and it is only thereafter that models may be abstracted and the technological imitation of nature carried out.

The above analyses also presented a view of the human place in nature that is of fundamental significance to the discussion of biomimetic ethics in chapter 3. According to the doctrine of enlightened naturalism, we humans are not just another type of living being characterized by very high levels of intelligence; as the sole inhabitants of the clearing, we are ontologically different from all other organisms in that we alone can understand what and that beings "are." And yet we nevertheless share with these other organisms the common trait of belonging to Gaia, of inhabiting the earth. The clearing, we may conclude, is a space that has opened up within Gaia, and that, as such, belongs to Gaia, and, because of its belonging to Gaia, what goes on in the clearing, what we make and what we do within the clearing, must respect Gaia's standards.[106] This is what it means to take "nature as measure": to apply Gaia's ecological standards within the clearing, the realm of humans.

With respect to chapter 4, the principal contribution made above has been the theorization of the concept of legein, for legein, as will later become apparent, is the process responsible for the production of knowledge in nature and, as such, underlies the possibility of learning from nature and thus also taking "nature as mentor"—the basic principle of biomimetic epistemology. It follows that, while we may in theory *imitate* purely physical beings, we can never truly *learn from* these beings; ultimately, it is only from beings in which legein is present—organisms, societies, species, and Gaia—that knowledge may be acquired.

## CHAPTER 2

# Nature as Model

## Biomimetic Technics

The biomimicry revolution is an epochal shift, a shift away from an epoch that largely ignored the imitation of nature to one that puts it center stage. This raises a question that has been surprisingly little addressed by biomimicry's leading exponents: If biomimicry is a process of technological innovation that takes nature as its model, how does it differ, exactly, from the technological innovation that has been taking place in the prior epoch? A revolution, after all, is not only the beginning of a new epoch but also the end of a previous one; understanding this previous epoch is therefore an essential element in understanding the revolution itself.

The idea of a biomimicry revolution also raises the question of its novelty. Benyus and others typically present biomimicry as something radically new, an almost unprecedented phenomenon in human history. And yet, as I noted in the introduction, a genuine revolution is not simply a radical rupture with the present; it is also a revolving-back to a certain moment or period in the past. This revolving-back is not, however, an identical repetition. It is a creative repetition, which, as such, may be seen as what Heidegger calls a "new beginning": a beginning that, while in some respects a return to the first beginning, is in other respects very different.[1] In addition, then, to the question of understanding the epoch with respect to which the biomimicry revolution constitutes a radical break, there is the more primary question with which this chapter shall begin: What is the first beginning that the biomimicry revolution must in some way or other renew?

In Heidegger, the concept of a new or other beginning applies to the thinking of being as physis as it first emerged with the early Greeks. Drawing on Heidegger, I argued in the previous chapter that the thinking of being continued thereafter, albeit no longer as physis, but in the post-Parmenidean guises of materialism and idealism. I further argued that the present epoch is on the cusp of the ontological revolution glimpsed by Heidegger, for the thinking of being as physis, which has resurfaced in the roots of cybernetics and systems theory (as anticipated by Heidegger), may be developed in such a way as to provide the philosophical ground of "another beginning."

If the biomimicry revolution is to come to pass, something similar, I contend, must also occur with respect to technē. Two general features of the concept of technē are of particular significance here. First, and most obviously, it was widely assumed among the ancient Greeks that technē arises from the imitation (mimesis) of nature (physis).[2] So, whereas imitating nature today appears as an unorthodox approach to technological innovation, for the ancient Greeks it was an essential aspect of technē. Second, the word *technē* refers to human making in general and, as such, covers both art and technology. This is not to deny that the ancient Greeks were unaware of any differences between the two, including differences between the production of functional and aesthetic artifacts. As we will see in later discussions of Plato and Aristotle, distinctions between art (e.g., painting, poetry) and technology (e.g., carpentry, pottery) were both commonplace and significant in ancient Greek thinking. Nevertheless, the fact that art and technology were covered by a single term is potentially enlightening, for it not only suggests the possibility of bringing art and technology together within a single area of philosophical inquiry (technics) but it may also facilitate the transfer of philosophical theories and analyses—of mimesis, for example—from one to the other.

It is with a view to reflecting these two points—seeing technology as essentially involving the imitation of nature and as belonging to the same field of philosophical inquiry as art—that I refer to the general field of inquiry of the present chapter as "biomimetic technics," where the word "technics" is to be understood in a sense that echoes the ancient Greek concept of technē inasmuch as it covers the study of both art and technology, and the word "biomimetic" as explicitly emphasizing the focus of this study on nature imitation.[3] But, before we move on to the task of presenting and exploring biomimetic technics, let us first consider how it was that we arrived at the contemporary situation in which the ancient Greek view of technē was lost and replaced with the altogether different view that holds sway today.

There can be little doubt that the ancient view of technē as imitation of nature survived the Parmenidean revolution in ontology. Both Plato and Democritus upheld this view, though, as a result of their radically different ontologies, their interpretations of technē were quite different. According to Plato, the genuine beings are the Forms or Ideas, with the sensible entities we ordinarily perceive being but degraded copies or imitations of these Ideas. In a similar vein, Plato thought that art (e.g., painting, sculpture, poetry) was an inevitably degraded imitation of the everyday things we commonly perceive via the senses. Art, he famously concluded, is but an imitation of an imitation, and thus at a "third remove from reality."[4]

Plato's conception of technology—understood here as the functional artifacts produced by technē—is less clear. In the *Republic*, he contrasts the craftsperson with the artist, arguing that the craftsperson directly accesses the Idea, which they proceed to reproduce materially, in which case only art, not technology, is imitation of nature, where by "nature" I mean the world of appearances.[5] But, according to Aristotle, Plato's Academy denied the existence of Forms for artificial objects.[6] Why this is the case becomes clear in the *Timaeus*: the demiurge realized the whole panoply of Forms, leaving none leftover for humans to imitate directly in technology.[7] This suggests that Plato ultimately concurred with the view that technological artifacts, like works of art, are but imitations of nature.

Democritus also viewed technē as imitation of nature. The true beings, he thought, are the Atoms. These Atoms, he further thought, combine to form the various natural things we encounter in the world of sensory experience, and it is these natural things that we imitate in technē. Examples provided are houses, which Democritus thought arose from imitating swallows' nests, and weaving, which he believed arose from imitating spiders' webs.[8] But if Democritus's thinking about technē qua imitation of nature differs from that of Plato in that it focuses on technology rather than art, it nevertheless resembles Plato's in the way that the imitations of technē are located at a third remove from reality. Far from imitating the genuine beings, the Atoms, works of technē imitate only "unreal" compositions of Atoms.

With the rise of Christianity to a position of ideological dominance in the Middle Ages, and therewith also the conception of nature as the creation ex nihilo of God, there arose the first significant challenge to the ancient Greek view of technē as nature imitation. The ancient view was not explicitly rejected, but it was given a different foundation. To imitate nature was no longer to imitate physis, or, even, post-Parmenides, to imitate the natural world as it derived from the Atoms or Ideas, but rather to imitate the creation ex nihilo of an omniscient and omnipotent God.

In some cases, moreover, the imitation in question was not thought to be of God's creation, but of the creative process at work in His creation. When Thomas Aquinas, for example, says that "art is the imitation of nature in her manner of operation," what he indicates was ultimately being imitated was the divine intellect at work in nature, the creative intelligence of God.[9] Going one step further, Nicholas of Cusa's character, the *idiota*, an artisanal spoon-carver, claims that what technology imitates is not God's creation, nature, or even the divine intellect at work in nature, but rather God's *act* of creation ex nihilo, or the "infinite art" as he calls it, which is what first creates nature.[10] It is important to realize, however, that the idiota remained an idiota; his view was held in private (*idion*) and not yet accepted in public (*koinon*). In public, the ancient Greek view of art as imitation of nature was upheld, albeit on the basis of different theoretical grounds from before. Nature imitation remained the norm, even as the imitation of God—theomimicry—now provided its ultimate justification and ground.

As the Middle Ages drew to a close, theomimicry—the view of God as model—was replaced by anthropomimicry, the view of Man as model. This shift is visible above all in the architects of the early Italian Renaissance, including Leon Battista Alberti, Filarete, and Francisco di Giorgio Martini.[11] The thinking of the latter is particularly representative of Renaissance anthropomimicry. Starting from the observation that other species always build the same things (e.g., nests, webs) in essentially the same way, as determined by their nature, Francisco notes that humans alone are free to build things in ways of their choosing. This in turn raises the question of *how* they should build. In answering this question, he reasons that since the human being is the most perfect in all nature, it should provide the model for art.[12] The ancient concept of art as imitation of nature could thus be explicitly upheld while at the same time entirely reduced to the imitation of human beings. Even when the Italian Renaissance architects proposed designs that bore little obvious resemblance to human beings, a human model characteristically lurked underneath. The figures of the circle and the square, for example, were regularly employed on the grounds that they share a profound geometrical affinity with the human form, as is manifest in that most famous emblem of Renaissance art, Leonardo Da Vinci's *Vitruvian Man*.

Anthropomimicry was also present at the inception of modern political philosophy in Thomas Hobbes's *Leviathan*. Beginning by remarking that art is the imitation of nature, Hobbes goes on to claim that the human body—the "most excellent" thing in all of nature—provides the model for the state, which he thus conceives as an "artificial man."[13] When one further considers Hobbes's view that subsequent to the establishment of the social contract the totality of human activity participates in the activity of the state, it follows that all human activity beyond the state of nature is in some sense also anthropomimetic. As in the case of Francisco di Giorgio Martini, the ancient Greek view of technē as imitation of nature is thus explicitly upheld, even if it is only ever the human being that is imitated. Further, the human being is not just what Aristotelians call the "formal cause" of the state, but also its "material," "final," and "efficient" causes; human beings provide the *form* of the state, the *material* of which it is composed, the *end* it serves, and the *means* by which it exists. The only breach in this wholly anthropomimetic conception of the state concerns its scale, for Hobbes's "artificial man" was of "greater stature and strength than the natural," which is why he has recourse to the mythical figure of the leviathan, a giant sea monster.[14]

If medieval theomimicry and modern anthropomimicry provided the ancient view of technē as imitation of nature with different foundations, from the early nineteenth century onward this ancient view came under sustained attack. The pivotal moment is probably the thought of G. W. F. Hegel. According to Hegel, the "most common opinion" about art holds that it aims to imitate nature.[15] On this standard account, art succeeds when it reproduces as faithfully as possible something that already exists in the external world. Hegel criticizes this common opinion in two ways. First, he argues that to see art as imitation of nature implies that it is superfluous, for if art aims only at faithful imitation one would be better off directly observing things as they already exist in nature.[16] Second, Hegel argues that the view of art as imitation of nature manifestly fails to describe many instances of art. Architecture, he tells us, is not an imitation of anything. Likewise, since poetry is not purely descriptive, it is also, Hegel thinks, not an imitation of nature.[17] This is not to deny that, in the case of some arts, notably painting or sculpture, the artist could set out simply to imitate nature as faithfully as possible, but what they would thereby have produced, Hegel thinks, would not be a genuine work of art at all, but a purely mechanical accomplishment devoid of any meaning or value. It follows, Hegel claims, that any genuine human invention is more meaningful and valuable than any imitation, and even the most insignificant of technical inventions—he gives the examples of hammers and nails—of greater meaning and value than the most accomplished of imitations.[18]

This rejection of technē as imitation of nature also characterized the emergence and development of the first explicit "philosophies of technology" in the late nineteenth and early twentieth centuries. The expression "philosophy of technology" originated with Ernst Kapp.[19] Kapp's philosophy of technology centers on his concept of organ projection, which is comprised of two parts, one concerning function, the other form. The basic function of technological artifacts, Kapp claims, is to extend the reach and power of the human body. As for their form, Kapp thinks they all originate in what he calls the "morphological replication" of the various organs of the human body, as well as, in some cases, of the human body as a whole.[20] The hammer replicates the form of the forearm and fist, the spade the arm and outstretched palm, musical instruments the ear, optical instruments the eyes, railways the sanguine system, and the state the human organism as a whole. But while this may at first appear to be a straightforward continuation of modern anthropomimicry, it differs in one fundamental respect. Drawing on the newly forged concept of the unconscious, Kapp argues that morphological replication is not in any way conscious or deliberate; it is only post hoc that we may come to realize that our technologies reproduce the basic forms and functions of the human body.

Other philosophers of technology went further, denying even unconscious replication. The late nineteenth-century German philosopher of technology Max Eyth developed a theory of the "spiritual autonomy" of technology, arguing that, far from imitating nature, technological innovations should be seen as emerging from the "pure life of spirit."[21] A comparable position was put forward by the early twentieth-century German philosopher Friedrich Dessauer. In the act of technological creation, Dessauer claims, one accesses a quasi-Platonic realm of technological ideas, a realm containing "pre-established ideal solution[s]" to technical problems, including the basic technical solutions underlying such inventions as bicycles, planes, levers, saws, and taps.[22]

Dessauer also thought that this realm of technological ideas was radically "alien" to nature. The invention of successful technologies, he believed, requires one to *abandon* natural models and instead to discover technological ideas not based on nature at all:

Human flight differs completely from the flight of birds; it succeeded only when moving wings were abandoned. The sewing machine sews differently than man does; the mill grinds differently than teeth do; transportation takes place by means of wheels, not through leverage of legs. Thus many works of technology are built not by approximating nature but according to an order alien to nature. Where nature enters as inventor, producing ever new forms in the realm of organic life, it again follows an order completely inappropriate for technology.[23]

In keeping with this, Dessauer thought that, far from deriving from knowledge embodied in nature, technology derives from a "higher" realm of spirit, whose role is to organize and give form to the "chaotic" realm of material nature: "From a higher sphere of power and reality, through the spirit and hands of the technician and worker, an immense stream of experience and power descends into earthly existence. A spiritual stream pours into the chaotic material world, and everyone, from the creator to the final worker, takes part: all are recipients."[24]

Surveying the philosophical conceptions of technology put forward by the likes of Eyth and Dessauer, Ernst Cassirer claimed in 1930 that the rejection of nature as a model for technology may be considered nothing short of the "basic principle" of modern mechanical engineering:

> What separates the instruments of fully developed technology from primitive tools is that they have, so to speak, detached and dissociated themselves from the model that nature is able immediately to offer them. What these instruments have to say and accomplish—their independent sense and autonomous functioning—completely comes to light only because of this "dissociating." As to the basic principle that rules over the entire development of modern mechanical engineering, it has been pointed out that the general situation of machines is such that they no longer seek to imitate the work of the hand or nature but instead seek to carry out tasks with their own authentic means, which are often completely different from natural means. Technology first attained its own ability to speak for itself by means of this principle and its ever sharper implementation. It now erects a new order that is grounded not on contact with nature but rather, not infrequently, in conscious opposition to it.[25]

In a similar vein, the contemporary German philosopher Hans Blumenberg contended that from approximately the eighteenth century onward nature not only lost its status as model or exemplar to be followed, but even came to be seen as "antithetical" to successful technological innovation.[26] This in turn, Blumenberg noted, had the significant epistemological consequence that nature's utility was from this point on seen purely in terms of the materials and energy it provided, with the information or knowledge required for the invention of new technologies coming exclusively from humans, thus giving rise to a "sphere of pure construction and synthesis."[27] Technē, Blumenberg concluded, was no longer viewed as nature imitation, but as human creation.

As for more recent philosophy of technology, it has for the most part overlooked the question of the origin of technology. Heidegger, for example, follows Aristotle in seeing technology as arising from the four causes.[28] But he overlooks the question of where these causes come from. Discussing the production of a silver chalice, for example, he says that the form of the chalice is simply "chaliceness," but without telling us where chaliceness comes from. So, whereas the ancient Greeks would presumably say it comes from observing and imitating nature, perhaps the form of rock pools or water-collecting plants (e.g., bromeliads); Kapp that it comes from unconsciously replicating the form of a pair of cupped hands; Dessauer from discovering a preexisting idea or ideal solution to the problem of how to hold liquid for drinking; and Blumenberg (or, rather, the position he describes) from creating an entirely new form in the human mind, Heidegger has nothing at all to say on this matter.[29] Further, in keeping with this apparent indifference to the question of where the causes of technē come from, Heidegger's famous discussion of technology at no point even mentions Aristotle's own answer to this question: that technē arises through imitating nature.

This is not to say that the concept of imitation has been entirely abandoned. In the form of postmodernism, the "pantomimetic" idea has emerged that anything and everything may be taken as a model or source of inspiration for human technē. Every style, every epoch, every previous artist, every object, every theme—everything, in short—may henceforth come to act as model or source of inspiration, and these models may be freely combined or juxtaposed in provocative, unusual, and even contradictory ways.[30] But, while in postmodernism, anything and everything is a potential model, it is also true that these models tend to concern only the aesthetic features of the work, not, at least where doing so would be possible (e.g., architecture), its functional features. This, then, is why postmodernism has thus far been restricted to art.

If biomimicry requires us to break with the common view of technē as human creation, but also with the postmodern view of technē as the imitation of anything and everything, and instead creatively to renew the ancient view of technē as imitation of nature, the question arises as to how we are to go about and theorize this renewal. One obvious possibility would be simply to take the most general traits present in the ancient view and then to articulate them with contemporary thinking about biomimicry. Just as in the previous chapter I articulated the early Greek concept of physis as poiēsis en heautōi with contemporary thinking about autopoiesis and self-production, so ancient Greek thinking about technē as imitation of nature could be articulated with contemporary thinking about biomimicry.

While such an approach is no doubt necessary, it leaves us with relatively little to go on. If all we are taking from the ancient Greeks is the concept of technē as nature imitation and the idea of in some sense bringing together the study of art and technology under the rubric of "technics," that gives us little guidance as to how our analyses are actually to proceed. With a view to overcoming this problem, I will in the rest of this chapter draw not so much on ancient Greek thinking in general, but more specifically on the work of Aristotle, especially the *Physics* and the *Poetics*, which between them will provide much of the philosophical basis for the following discussions of biomimetic technics.

## ABSTRACTING NATURAL MODELS

### *The Impossibility of Imitating Self-Production*

If one is to imitate nature, nature must necessarily assume the status of model. But the principle of "nature as model" is not as straightforward as it may at first seem. For a start, it is important to realize that, when one imitates nature, it is not nature *as such* that is being imitated. Indeed, it is in fact a logical impossibility to imitate nature, where nature is understood as physis—that is to say, as self-production. Anything produced by humans is by definition an instance of technē; it is produced by something other than itself, even when based on a natural model. Self-producing beings, it follows, cannot be imitated as such—that is, inasmuch as they are self-producing. That would involve the production of something that produces itself by something other than itself, which is a logical contradiction. One cannot produce physis by way of technē; one cannot produce a natural being artificially.

A first important consequence of this concerns the contemporary project of creating life artificially, which is sometimes assumed to fall within the scope of biomimicry.[31] According to Maturana and Varela, there is nothing contradictory about the idea that humans may one day fabricate self-producing or autopoietic beings.[32] Current attempts to achieve this end, such as the technique of inserting an entirely artificial genome into a living cell, would no doubt be considered to have fallen short; what is created in this instance is only the genome, not the entire cell of which the genome is a part. But there is, Maturana and Varela maintain, nothing contradictory about artificially producing a living, self-producing being.

In light of the logical contradiction identified above, the mistake I believe Maturana and Varela are making here lies in a failure to distinguish between artificially producing a self-producing being, and artificially producing the conditions in which a being may produce itself. There is certainly nothing logically contradictory about the idea that a living being may emerge de novo in a laboratory, but if that were one day to occur it would not mean that we ourselves had created life—that a self-producing entity had been produced by something other than itself, which is a logical contradiction—but that we had succeeded in producing the appropriate physicochemical conditions (e.g., macromolecular structures, nutrient resources, temperatures) required for a living being to produce itself. The quest for artificial life, then, is ultimately just the quest to produce conditions conducive to living beings bringing themselves into existence de novo. The same is true, moreover, of all self-producing beings. Even vortices, which are not generally considered difficult to produce artificially, only ever produce themselves. We may easily be able to create the right conditions for vortices to self-produce (e.g., by removing the plug from a full bathtub), but we do not produce or make the vortex itself.

But the negative result that one cannot artificially reproduce nature itself, in the sense of physis, leaves open the question of what imitating nature does involve. What does it mean to take nature as model? And what are the different elements and steps involved in that process?

## A Typology of Imitable Beings

While it is not possible to imitate the self-production of physei-beings, it does not follow from this that there is no sense at all in which physei-beings may be imitated. They may be imitated not inasmuch as they are self-producing, but inasmuch as they possess traits other than self-production. But, before we look at the imitation of the various traits that may be abstracted from physei-beings, let us first consider in more detail the various beings with respect to which the process of abstracting models may occur.

In keeping with the discussion of physis in the natural sciences set out in the previous chapter, these beings may be physical, biological, or ecological. The three principal instances of physical self-production I discussed were atoms, vortices, and stars, and it is significant that all of these have provided models for techno-logical innovation. The three-dimensional geometry of carbon atoms has been imi-tated by combining microscopic polystyrene spheres coated in DNA. These structures have in turn been joined to one another so as to imitate the geometry of certain molecules, such as methane. Being much larger than the natural atoms on which they are modeled, these imitation atoms and molecules may potentially be used in the development of ultrafast optical computers, for their dimensions match the wavelengths of light.[33] Likewise, it has long been known that humans may one day be able to imitate the process of nuclear fusion whereby energy is generated in stars, perhaps the key technological obstacle being that artificial fusion reactors cannot rely, as stars do, on gravity to produce a limit or bound-ary within which their internal reactions are confined and must in turn resort to artificial confinement techniques. Lastly, as we have already seen, the form of cer-tain vortices has been imitated in order to produce efficient water mixers; in producing themselves in such a way that their internal entropy is kept low, vor-tices "enduce" high levels of entropy (a measure of disorganization) in their sur-roundings, which is precisely what efficient water mixers are required to do.

It is important to realize, however, that there is much less scope for imitating purely physical autopoietic beings than biological or ecological ones. The basic reason for this is the absence of gathering or legein. Through legein, especially in the form of natural selection operating on heritable variations, a great number of different traits may come into and remain in existence (or, rather, ensistence). If these traits do remain, it is because they tend to allow the being to which they belong—whether an organism, a society, a species, or Gaia—to endure. And all of these traits, as well as the various different forms of legein that give rise to them, are possible objects of mimesis. In the case of purely physical beings, by contrast, the absence of legein means that they do not evolve, and so, unlike in the case of biological and ecological instances of autopoiesis, we cannot imitate the fruits of their evolution (for there aren't any).

Further, since any traits purely physical beings do possess have not been honed by natural selection operating in similar contexts to those in which human beings find themselves, it will be pure coincidence if they are useful to us as mimetic resources. The three-dimensional geometry of carbon atoms may be a useful object of mimesis on account of the fact that, when imitated in the form of three-dimensional polystyrene spheres, it may help us develop ultrafast optical computers, but this mimetic utility is a coincidence. By contrast, that natural photosynthesis provides a useful model for artificial photosynthesis is no coincidence at all; like plants, we today require a reliable and sustainable source of usable energy. In the vast majority of cases, therefore, the imitation of nature will be quite accurately described as "biomimicry," for it is primarily from living systems (organisms, societies, species, and Gaia) that imitable traits may be abstracted.

## A Typology of Imitable Systems

In the previous section, we briefly surveyed the various different types of self-producing beings from which models may be abstracted. In this and the following sections, our attention will turn instead to the process of abstracting natural models.

There exists a standard—indeed, standardized—way of characterizing the biomimetic process of taking nature as model, which is set forth in the ISO 18458, "Biomimetics—Terminology, Concepts, and Methodology."[34] According to this characterization, the biomimetic process consists of three steps: 1) the observation of a biological system; 2) the abstraction of a model from that system; and 3) the transfer of the model into the target technological system.

The concept of a biological system is defined here as a "coherent group of observable elements originating from the living world spanning from nanoscale to macroscale."[35] Apart from the obvious question of what exactly is meant by the expression "living world," and which entities belong to it, what remains unanswered in this definition is the question of who or what delimits and thus brings forth the biological system. Even when a definition is provided of a "system" as a "set of interacting or interdependent components forming an integrated whole with a defined boundary," the problem remains, for it is uncertain who or what is responsible for defining the boundary.[36] Is it the system itself or is it the human observer? In short, it is far from clear whether the biological systems in question are self-producing beings (which, as such, delimit themselves); whether they are parts of self-producing beings delimited and thus brought forth by us; or whether they could be either of these.

With a view to resolving this ambiguity, I propose a slightly different way of characterizing the initial stages of the biomimetic process. The first stage is simply *the observation of nature*, where by "nature" I mean self-producing beings. These self-producing beings may either be observed in relation to an external environment, as when one observes a kingfisher diving toward its prey, or internally, as when one observes, say, the kingfisher's internal organs, though combining the two is also possible. The second stage is *the delimitation of a natural system from within a self-producing being and the subsequent study of that system*. At first sight, it may seem strange that this stage involves two activities: delimitation and study. But the reason these activities belong together is that they are both constitutive elements of *science*. In keeping with its etymology, science involves cutting reality up into component parts (or elements). But science also involves studying these parts, both in isolation and interaction with other parts, an activity that will likely involve further observation, but that may also involve characterization, analysis, comparison, and the like.

An important consequence of introducing this second stage into the biomimetic process is that models are not abstracted directly from self-producing beings; a "natural system" must first be delimited for the purpose of scientific study, such that it is ultimately from that artificially delimited system that models are abstracted. Consider the case of the Shinkansen 500 series, a high-speed Japanese train modeled in part on the form of the kingfisher's beak.[37] Now, one does not directly abstract the form of the beak from the kingfisher qua self-producing whole; an intermediary process intervenes, whereby a part of the kingfisher is delimited from the rest—in this instance, its beak—and in such a way that it may be studied scientifically with a view to abstracting a model from it; in this instance, its form. The self-producing being does of course remain the distal or indirect source of natural models, but the proximal or direct source is a natural system artificially delimited by humans.[38]

The natural systems that may be delimited from self-producing beings are of two types. The first, which I call "local systems," are systems restricted to a specific region of a self-producing being. Examples drawn from biological individuals would include such systems as the beak of a kingfisher, the paws of a cat, or the chloroplasts of plants. As for biological societies, the various structures they produce also fall into the category of "local systems." Ecological instances of local systems would include the various ecosystems, biotic communities, and the like, that it is possible to delimit from within Gaia, and at various different scales, such as a tundra biome, a conifer forest, or a rotting tree trunk.

The second type of natural system we may delimit from self-producing beings are what I call "global systems." In contradistinction to local systems, global systems are spread across more or less the entirety of the being in question, but without coinciding with the self-producing being itself. Biological examples would include the nervous system or the immune system, as well as the communication systems responsible for swarm intelligence, in the case of some societies. In the case of biological species, examples of global systems include the evolutionary interactions of genetic diversity and natural selection of fitness-enhancing alleles that enable species to adapt to changing environmental circumstances. Lastly, ecological examples would include various systems operative at the global scale of Gaia, such as the system of interactions between the oceans, atmosphere, and the biota responsible for global climate regulation, or the biogeochemical cycles (e.g., of water, carbon).

## A Typology of Imitable Traits

Like the process of delimiting natural systems, the abstraction of imitable traits from these systems is something carried out by humans. Any imitable trait we abstract is therefore artificial—or, rather, partly artificial, in the sense that it is abstracted by us, not by itself or by the self-producing entity in which it previously "ensisted." But artificiality does not necessarily entail arbitrariness; a typology of imitable traits may be more or less appropriate to the practice of biomimicry.

One way in which we might seek to answer the question of what an appropriate typology might be is through investigating the existing literature on the subject, as it is here that biomimicry researchers have set out the typologies that they themselves consider appropriate. Without presenting a detailed literature review, the approach generally adopted consists in selecting a restricted number of traits, often between three and six, usually from among the following: forms, structures, organizations, materials, processes, mechanisms, functions, behaviors, constructions, and systems.[39] While usefully indicative of the sort of things one might find in an appropriate typology of imitable traits, from a philosophical perspective these typologies are unsatisfactory, for it is usually not clear on what basis they have been established. This raises the question of how an appropriate typology might be established.

There are, I suggest, three criteria that an appropriate typology should satisfy. First, it should be *economical*, in the sense of consisting of the minimum number of basic types of natural models. One could not, for example, include wood, nacre, and bone in such a typology, for these all belong to the more basic category of materials. Second, it should be *comprehensive*, by which I mean that it should cover all the different types of basic natural models we may seek to imitate. And, third, it should be *cohesive*, in that the basic types it advances should fit together, without significant overlap, like the pieces of a jigsaw. But while these criteria constrain any typology we might put forward, they do not in themselves suggest any particular typology.

In view of this, I shall, like many of my colleagues, begin by simply putting forward a typology. My typology differs from theirs, however, in two main ways. First, it is not "mine" in the sense that it was not invented by me; it derives instead from Aristotle's conception of natural science, and more specifically from his theory that everything may be understood in terms of four causes, traditionally referred to as the formal, material, efficient, and final causes. Second, I will also offer explicit justification of it. Since, however, the typology derives from Aristotle, this approximates to a justification of the claim that an appropriate typology may be derived from Aristotle.

A first reason for turning to Aristotle here relates to the resulting typology's overlap with existing literature on the subject. The concepts of matter and form—to take only the two most obvious examples—are found in both Aristotle's theory of the four causes and, as far as I can tell, in all contemporary typologies of natural models. But this similarity does not simply suggest a certain overlap and compatibility with contemporary thinking about biomimicry. Indeed, the basic distinction made today between form and matter, and not just in the context of biomimicry, is in large part a legacy of Aristotle, who developed the influential doctrine of hylomorphism, according to which every being is a compound of matter and form, in response to Plato and Democritus. This suggests that contemporary typologies of imitable traits may ultimately derive, at least in certain key respects, from Aristotelian natural science, which in turn suggests that it is ultimately to Aristotle that we should return in trying to work out an appropriately "revolutionary" typology of imitable traits.

A second reason is the general pertinence for biomimicry of Aristotle's conception of natural science. Plato, as we have already seen, thought that prior to and underlying the material beings we perceive with the senses are the Forms or Ideas, which are accessible only to reason. From Aristotle's empiricist perspective, by contrast, forms only come into existence as *abstractions* from concrete

individuals.[40] In this respect, his thinking ties in neatly with the equally empiricist view present in the standardized interpretation of the biomimetic process
as involving the *abstraction* of natural models. Importantly, however, Aristotle
does not limit the types of traits that could be abstracted from nature to the
forms. As the doctrine of the four causes implies, he thinks it is also possible to
abstract other traits from any given natural being—namely, the material (or
materials) of which it is composed, the function (or functions) it accomplishes,
and the generative process (or processes) that brings it into being, and which
together, he thinks, "cause" the being in question. It is then but a short step—
albeit one not taken by Aristotle himself—to seeing these four causes as also
providing the basis of a typology of traits that may be imitated in human design.

The third reason Aristotle's theory of natural science may provide the basis
for an appropriate typology of imitable traits concerns the fact that it applies to
both nature and technology. One may talk, for example, about natural materials, just as one may talk about human-made, artificial, or synthetic materials. This
dual applicability makes possible a seamless transfer of causes, in the Aristotelian sense, from natural entities to artificial ones; a natural material may provide
a model for an artificial one. If, on the other hand, nature and technology called
for analysis in terms of different sets of causes, or if it were impossible to analyze
one or the other in terms of multiple distinct causes, then any transfer from one
to the other would be rendered problematic or impossible.

This last point provides the occasion to discuss an important feature or implication of Aristotle's theory of natural science. It is only to the natural systems
delimited by natural science, and not to self-producing beings as such, that Aristotle's fourfold conceptual schema applies, for, in the case of self-producing beings,
viewed as such, it is not possible to separate out the different causes from one
another. Just as contemporary physics tells us that the four fundamental forces
were unified at the very early stages of the universe, so the thinking of physis tells
us that the four fundamental causes are unified in the case of self-producing beings
and only become differentiated from one another subsequent to the artificial
delimitation of natural systems.

With a view to substantiating this claim, let us first examine the formal and
efficient causes. It is not hard to see that in the case of physis these causes are the
same, for self-production is both the form of self-producing beings and the generative process which brings them into being. When Maturana and Varela affirm,
for example, that self-production is the *organization* of the living, they mean

something very similar to the claim that it is the *form* of the living. But self-production is not only the form or organization of the living; it is also the generative process that brings the living being into being.[41]

But what, then, of the final cause? According to Freya Mathews, living beings, which she explicitly conceives as self-producing or autopoietic, have an end, or telos, which is to self-produce.[42] On this view, it is not just the formal and efficient causes that are identical, but also the final cause. The notion of an end or telos is, however, widely thought to be problematic when applied to living beings, and the same point no doubt holds also—and even a fortiori—in the case of other autopoietic entities, especially purely physical ones. But if one replaces the notion of end or telos with that of function, defined nonteleologically, this problem may be overcome. Defined in this way, the concept of function refers simply to "what something does." What a car does, and thus also what its function is (or at least one of its functions), is to transport people from one place to another. Likewise, what a leaf does, and thus also what its function is (or at least one of its functions), is to produce usable energy from the sun. But what a self-producing being does, and thus also what its function is, is simply to produce itself.[43]

Of course, a self-producing being may also do all sorts of other things. A star, for example, may give off light and create heavier nuclei in its core. That is not, however, to see it any more as a *self-producing* being, but as a being that produces—or, rather, enduces—things other than itself. By contrast, inasmuch as stars, living beings, and species are conceived as self-producing, self-production is also what they do. The word "self-production," it follows, describes not only what self-producing beings are and how they come into existence but also what it is that they do—that is to say, their function; they produce themselves.

Lastly, let us consider the material cause of self-producing beings. Now, it may at first seem as though this cause alone is quite separate from the others. While self-production is the formal, efficient, and even, mutatis mutandis, the final cause of self-producing beings, it is not the material cause; self-producing beings are not composed of self-production, but of various materials. This is of course true, but, as I will now show, it is also true that, while not *identical* to the other causes, the material cause is nevertheless *inseparable* from them.

In order to understand this claim, let us begin by considering two classic Aristotelian examples of technological artifacts: the silver chalice and the bronze statue. There can be little doubt that form and matter are separable here; Aristotle could just as well have talked about a bronze chalice and a silver statue. It's not that Aristotle does not recognize matter as *relative* to form; part of the skill of the artisan or artist, he thinks, lies precisely in selecting a material that is appropriate for the form in question (e.g., bronze for one statue, marble for another).[44] But even here the two causes remain separable and are only contingently conjoined in the end product.

In the case of self-producing beings, by contrast, the material cause is inseparable from the others. Human beings, for example, are necessarily composed of various types of cell (e.g., muscle, nerve, skin), along with other organic and inorganic substances, and could never be composed of different materials from these, minor artificial prostheses or additions notwithstanding. So, while it is of course possible to abstract a form from a system that has already been delimited from within a human being, such as the form of the hands, and then to reproduce that form in a variety of different materials (e.g., plastics, metals), nothing like this separability of form and matter obtains in the case of the human beings themselves, understood as self-producing beings.

All of this points to the following conclusion: Aristotle's doctrine of the four causes is not applicable to self-producing beings *as such*, because in the case of self-producing beings the four causes are, if not always identical, then at least inseparable from one another, and, as such, cannot be separated from one another by means of abstraction. Therefore, if one wants to understand that to which self-producing beings owe their existence (i.e., themselves, as self-producing beings), one must turn not to natural science, understood here in Aristotle's sense as the doctrine of the four causes, but rather to natural ontology, understood as the thinking of being itself as physis. If, however, one's aim is not to understand physis, but to analyze natural systems abstracted from self-producing beings for the purposes of imitation, then a fourfold typology derived from Aristotle is of great value, for, since it may be applied to both natural and technological systems, traits abstracted from the former may be unproblematically transferred over to the latter.

### Levels of Abstraction

The final important concept to consider in this context is what I call "levels of abstraction." The need for this concept arises from the fact that it is possible not only to abstract traits from physei-beings (via the delimitation of natural systems) but also to abstract traits from these abstracted traits, thereby progressing to ever higher levels of abstraction. There are, in short, not just abstractions (level 1), but also abstractions of abstractions (level 2), abstractions of abstractions of abstractions (level 3), and so on.

The importance of these higher levels of abstraction is that they make it possible to abstract not just specific traits from this or that specific being, but general traits potentially common to a number of beings. An example may prove instructive. It is well known that one may imitate the microscopic form of the surface of the lotus leaf, which, precisely because it is not smooth but punctuated by a number of protrusions, prevents water droplets from sticking to it and instead causes them to roll off, taking dirt particles with them. But from this level 1 abstraction one may abstract the level 2 model of microscopic surface structures that prevent the accumulation of dirt, but do not necessarily have exactly the same form as the surface of the lotus. From this level 2 model, one could then abstract the level 3 model of increasing information content in order to reduce energy expenditure, for whereas a smooth surface contains little information but may require regularly cleaning and thus energy expenditure, modifying the surface texture in ways that prevent the accumulation of dirt increases the information content of the surface while also reducing energy expenditure.

In these instances of higher-level imitation, it will often be more appropriate to talk not of the abstraction of "models," but of the abstraction of "principles." The level 3 model of increasing information content to reduce energy expenditure may thus be referred to as an abstract "principle" of biomimicry. It is also important to realize that, precisely because higher-level traits may often be abstracted from a wider range of natural beings than lower-level traits, while also covering a much wider range of specific instances of nature imitation, they may additionally function as higher-level categories under which the imitation of lower-level abstractions may be subsumed. Concrete imitations of the microscopic surface structures of, for instance, lotus plants and shark skin could be categorized as imitations of the abstract level 2 trait of self-cleaning microscopic surface structures, thus giving rise to such abstract, higher-level research fields as "biomimetic self-cleaning surfaces" under which the relevant level 1 instances may be subsumed.

## PERMUTATIONS OF NATURAL MODELS

### Imitating Natural Systems

In the preceding analyses, I provided answers to four questions: first, the question of the *types of beings* from which imitable traits may be indirectly abstracted—namely, self-producing beings, whether physical (atoms, vortices, stars), biological (individuals, societies, species), or ecological (Gaia); second, the question of the *types of systems* that we may delimit from within self-producing beings and from which imitable traits may be directly abstracted (i.e., local and global systems); third, the question of the *types of imitable traits* that may be abstracted from these systems (i.e., forms, materials, generative processes, and functions); and, fourth, the question of the *levels of abstraction* of imitable traits— that is, level 1 (abstractions), level 2 (abstractions of abstractions), and so on.

In what follows, I present an overview of the various permutations of the abstraction process by means of a number of both real and hypothetical examples of biomimetic innovation. In each case, the abstraction is at a certain level, of a certain type of trait, and from a certain type of natural system delimited from within a certain type of self-producing being (e.g., a level 1 abstraction of a form from a local system delimited from within a living being). This overview will be structured in the first instance by means of the fourfold Aristotelian typology of imitable traits, and for three reasons: 1) because doing so focuses attention on the process of abstraction itself (as opposed to its level or the type of being or system to which it applies); 2) because it provides the opportunity for further discussion and illustration of these four types of traits; and 3) because doing so provides a concrete opportunity to show—albeit only in an indicative and a posteriori manner—that this typology does indeed respect the three criteria identified above as necessary for it to count as appropriate: economy, comprehensiveness, and cohesion.

Before we examine these four types of traits, there is a prior issue that must be addressed: the question of whether one may *directly* imitate natural systems. In the preceding discussion, I claimed that one first delimits a natural system and then abstracts imitable traits or models from that system. Systems, from this perspective, cannot simply be placed on a par with such imitable traits as forms and materials. But the question remains as to whether one may imitate natural systems directly, without passing through an intermediary process of abstracting imitable traits. Consider the case of a simple natural system, such as a protein. Now, there can be little doubt that one may imitate it in every respect—its materials, forms, generative processes, and functions. But can one imitate the protein without distinguishing between these four different traits? Can it be the protein qua concrete natural system that one imitates, rather than the protein qua complex of four imitable traits?

With a view to answering this question, let us consider the production of the artificial, biomimetic protein. In order to produce this protein, one must identify and gather together the requisite materials, while thinking about the form into which these materials are to be arranged, what function they are to have, and how it is that the protein is to be generated. One cannot, it would seem, think otherwise about the production of the artificial protein than in terms of the four-fold conceptual schema derived from Aristotle. But, since the artificial protein arises through imitation, one must also identify the materials, forms, generative processes, and functions of the natural system. When one thinks, as one must, about the materials from which the artificial protein is to be made, one cannot but think, too, about the materials of which the natural protein is made (at least if one is imitating the protein in toto as opposed to only, say, its form).

From this perspective, it is certainly possible to imitate natural systems, but the natural system is necessarily understood here as a complex of imitable traits, as some sort of combination of materials, forms, generative processes, and functions, and not as a concrete entity in which these traits remain undifferentiated. This explains why systems will not be treated alongside forms, materials, generative processes, and functions as another type of natural model.

## Forms

Much of the work carried out within biomimicry concerns the imitation of forms abstracted from nature. These forms come in two main varieties: structures and organizations. By "structures," I mean forms whose parts do not move significantly in relation to one another. By "organizations," I mean forms whose parts move in relation to one another, whether while remaining attached in predetermined ways (as in musculoskeletal systems) or while coming into contact only intermittently (as is typical in ecosystems).

Starting with structures, let us consider the example of a level 1 structure abstracted from a living being: the above-discussed form of the kingfisher's beak. On account of the way this structure allows the kingfisher to pass smoothly from one physical context to another (air to water), it provided a model for the Shinkansen 500 series, which, because it had frequently to pass in and out of tunnels, found itself in an analogous context to the kingfisher.[45] Another classic example of a level 1 structure abstracted from a living being is the microscopic surface of the lotus plant, which has provided a model for various self-cleaning surfaces, including paints, glasses, and textiles, and which, as we have also seen, is also a potentially important source of more abstract level 2 and 3 models.[46]

The imitation of structures very often concerns traits abstracted from biological individuals. There are, however, other types of beings from which structures may be abstracted. Obvious structures that may be abstracted from insect and animal societies relate to their collective constructions, which are important sources of models in architecture and urban design. These may also give rise to level 2 abstractions, such as the general trait of being adapted to their geographical context (e.g., prevailing winds, temperature changes), and that may in turn provide a guiding principle for the forms of human constructions.

As for Gaia, the dynamic nature of her component ecosystems means that there is probably less scope here than in the case of living beings or societies for abstracting structures for the purpose of technological imitation. This is not to say that Gaia affords no possibilities here. One of these consists in imitating the morphology of geographical features, most obviously in order to minimize disruption to the local biota or to the aesthetics of a site. Architects in particular may embed their structures into the landscape in such a way that they follow its natural contours, while also—at least in some cases—allowing the local biota to implant itself on or around the buildings (e.g., by placing a layer of soil on the roofs).[47]

The other major types of form that may be abstracted from nature are organizations. Obvious level 1 instances of organizations that may be abstracted from living beings relate to their various global systems, including sanguine systems, immune systems, and nervous systems. The organization of the human sanguine system, for example, has already functioned as a model in the design of urban water systems.[48] Organizations abstracted from the immune systems of vertebrates have taken as models for artificial immune systems, the basic goal of which is to imitate the abilities of learning and memory present in natural immune systems.[49] And nervous systems, including noncentralized ones (i.e., animals without brains), possess organizations that may be used as models in electronics and robotics.[50] Similarly, the way brains are organized furnishes important models for AI systems, in particular those studied under the rubric of "brain-inspired computing."[51] As regards biological societies, important level 1 models that may be abstracted here concern the organizations involved in the above-discussed cases of swarm intelligence, as present in ants, bees, and other species, and that may likewise furnish models for AI systems. These forms are also importance sources of level 2 models, including most obviously the general organizational principles of swarm intelligence.

Turning to ecology, one obvious source of level 1 organizations that may be abstracted from Gaia are forest ecosystems, including their trophic cycles; division into multiple layers (e.g., forest floor, shrub layer, understory, canopy, and overstory); or the complex networks of the rhizosphere, all of which may provide models in such fields as forestry (analog forestry);[52] agroecology (agroforestry, permaculture, and organic farming);[53] and urbanism (biomimetic or ecological urbanism).[54] Other organizations could be abstracted from prairies, which, thanks to the perennials of which they are mainly composed, maintain their characteristic organization even in winter (an important model for biomimetic cereal production),[55] and wetlands, the complex organization of which provides models for sewage treatment plants and artificial wetlands, often used also for the purpose of water purification.[56] Of course, in many of these cases, what will be imitated is not only the forms of natural ecosystems but also various materials, generative processes, and functions, but, in keeping with the discussion of the imitation of natural systems presented above, even in such a scenario it is necessary to conceptualize these traits separately. When establishing an agroforestry system, for example, one will need to think about both the plants one is to cultivate (i.e., the biotic materials) and how one is to arrange them (i.e., their organization).

Ecology is also a major source of level 2 organizational models. Such fields as agroecology, industrial ecology, ecological design, and ecological urbanism do not typically abstract level 1 organizational traits from nature for the purpose of imitation, but instead abstract level 2 traits common to a wide range of ecosystems. When, for example, systems designers abstract from nature such ecological principles as the circular recycling of nutrients or the dynamic hierarchical organization of trophic levels, level 2 organizations common to many concrete ecosystems are being taken as models. Further, it is important to note that these latter examples provide a powerful demonstration of how the concept of levels of abstraction allows for a significant expansion of the scope of biomimicry, such that it comes to cover more abstract levels of nature imitation, which have thus far typically been categorized as belonging to quite separate fields of design, such as agroecology, industrial ecology, and ecological engineering.

## Materials

In the textbook examples of technē provided by Aristotle, the distinction between form and matter is clear. In the case of the bronze statue, the form is that of a human being (in its outer shape), and the matter is the bronze. In the case of nature, however, any such simple distinction between form and matter is complicated by the fact that very often what, from one perspective, is matter is, from another perspective, form. Consider the human body. The matter of which it is composed includes living cells, blood, bone, and hair. But these materials are themselves composed of various molecules and atoms, which are in turn composed of elementary particles. Conversely, when we proceed in the opposite direction, toward the macroscopic, human beings appear as materials out of which human societies are composed, and these in turn appear as material components of Gaia.

For the doctrine of materialism, which identifies being with matter, this state of affairs is problematic. It makes being relative to human perspective, which is why materialists typically follow the ancient Greek atomists in seeing everything as composed ultimately of elementary units (e.g., the elementary particles of the standard model), which alone count as true matter, with everything else being more or less complex instances of form (and perhaps also emergent properties of these forms). From the present perspective, by contrast, matter is not being or nature, but rather something that arises (i.e., comes to exist) primarily through processes of delimitation and abstraction. The matter of a kingfisher's beak, for instance, arises through delimiting the beak qua natural system and then abstracting the matter of which it is composed. From this latter perspective, then, it is in the nature of matter to be relative to human perspective, for matter here is simply the relevant components of the system being analyzed. The bronze out of which the statue is made does not count as matter because it is irreducible to other more fundamental parts (bronze is, after all, an alloy composed mainly of copper and tin), but simply because it is that out of which the statue is made. Similarly, the living cells out of which nervous systems are composed do not count as matter because they are irreducible to more fundamental parts, but simply because they are the elemental components of these systems. These cells may in turn be composed of other elements, but that is to take the cells as forms and no longer as materials. This conception of matter as relative to perspective is of great importance for biomimicry, for, as we will now see, it allows for a wide range of ways in which the materials of physei-beings may be abstracted and taken as models in biomimetic design.

As it stands, almost all thinking about materials within biomimicry has focused on a subset of natural materials: composite materials generated by living beings. Some classic level 1 examples of composite materials of great interest to biomimicry include spider's silk, which is stronger than steel relative to its weight and provides models for ecologically benign biomimetic innovations in various fields;[57] bone, which is very light relative to its strength and durability, and provides models for biomimetic materials—most obviously in the medical field of bone tissue engineering, [58]but also as potential replacements for steel and cement in the construction industry;[59] and byssus, a substance used by molluscs to hold onto rocks and that is a source of models for sustainable adhesives.[60]

There are also a wide range of level 2 traits that may be abstracted from biologically generated composite materials. One of these is a principle found notably in nacre (mother of pearl) of inserting thin layers of proteins between layers of minerals, the function of which is to stop and deflect cracks.[61] So, whereas many artificial ceramics shatter immediately on impact, this level 2 principle allows for the development of much more durable ceramics. Abstracting further to level 3, one arrives at such principles as hierarchical organization and self-healing. Another higher-level model that may be abstracted through surveying multiple individuals concerns the limited range of basic polymers out of which nature constructs its more complex materials. While humans use over three hundred polymers plus metals to achieve the full range of functional materials currently used, nature uses far fewer.[62]

Nature does not just produce composite materials that we may take as models for artificial imitations. It also composes forms from out of these composite materials, in which case we may also look to imitate nature with respect to the composite materials out of which its forms are composed. From this perspective, imitating nature would extend beyond, say, making artificial wood based on the model of natural wood—an active field of biomimicry research—to the direct use of natural wood as a construction material.[63] Another level 1 instance of using the same composite materials as found in nature would be the use of natural fibers for biodegradable packaging. This in turn suggests an obvious level 2 trait that may be abstracted from the material traits of nature: the use of natural, biodegradable materials rather than synthetic, nonbiodegradable ones. The possibility of using the composite materials of which many natural forms are composed, and not just trying to produce artificial imitations of these materials, is important in extending the scope of biomimicry, for it allows it to cover much of what is today classified under the concept of the bio-based economy.

In keeping with the principle that what counts as a material depends on perspective, it is also important to note that living beings are not only made up of composite materials that they themselves generate in the course of their self-production but also of basic molecular and atomic substances readily available in the environment (e.g., in the soil, water, and air, as well as in the consumable bodies of other living beings). It is certainly possible to use these materials to produce much the same natural systems as may already be found in nature. Examples of this abound in the branch of organic chemistry known as organic synthesis, an important concern of which is the artificial synthesis of natural products (i.e., naturally occurring compounds), either from simple inorganic substances (total synthesis) or from more complex compounds extracted from living beings (semisynthesis).

But it is also possible to use the same molecular and atomic materials as nature to produce artificial forms not modeled on natural ones. Much synthetic biology may be biomimetic in this sense. Such things as designed proteins or industrial enzymes may in some cases not be modeled on natural forms—or at least not on any level 1 forms—and yet they typically use the same basic natural materials (e.g., amino acids). Further, the idea of using the same molecular and atomic building blocks as nature may furnish an important level 2 model for biomimicry. Just as nature uses far fewer base polymers than humans, so the vast majority of natural chemistry is carried out using only twenty-eight basic atomic materials, whereas we draw on almost the entire periodic table.[64]

To make things out of approximately the same basic atomic and molecular materials as nature could give rise to numerous benefits, including alleviating the dangers of resource depletion, facilitating recycling, reducing energy consumption, and creating significantly more opportunities for cooperating with other species—via nutrient exchange, for example—as opposed to accidentally harming and even destroying them with synthetic materials, including plastics and toxic chemicals. Once again, this greatly expands the scope of biomimicry, such that it could come to cover much of what takes place under the rubric of green or sustainable chemistry.

It is also important to realize that living beings are not only composed of atomic, molecular, and macromolecular substances of the sort studied by chemistry but also of living cells, and that it is possible to take these as materials for the construction of artifacts. A level 1 example is tissue engineering, which uses living cells as materials for artificial tissues, usually to replace or improve natural tissues, potentially in combination with abiotic biomimetic materials also developed in the context of tissue engineering.[65]

In addition to imitating the material traits of biological individuals, it is possible to imitate the material traits of the constructions of biological societies. One could, for example, seek to imitate mound-building termites with respect to both the forms of their constructions and the materials of which they are composed. Likewise, moving on to level 2, one could abstract the general principle of using locally available materials, as is also the case in much vernacular architecture, but not, generally speaking, of modern architecture.

The material traits of ecosystems furnish further important models for biomimetic design. Ecosystems, as we have seen, are composed of both biotic and abiotic materials. The biotic materials include both biological individuals and biological societies. The abiotic materials include such things as air, water, humus, sand, and rock. To imitate the material traits of an ecosystem would be to design and develop systems that are intentionally made of similar biotic and abiotic materials. Consider, for example, the idea of taking the indigenous forest as the model for a biomimetic city in the same location. To imitate the material traits of this system could involve using wood or other nonliving materials already present in the indigenous ecosystem for construction purposes. But, since the living beings themselves also count as material components of the system, it would also be possible to use them as materials for a biomimetic city. A biomimetic city based on the model of the indigenous forest could thus involve retaining or replanting a significant percentage of the indigenous trees, in which case the city would not only be composed of buildings based on the models of trees but also of trees themselves. Another example would be using the biotic and abiotic materials of a natural ecosystem as models in an agricultural system. This could involve such techniques as mixing crops and livestock, planting polycultures rather than monocultures, employing horses rather than tractors, attracting natural predators rather than spraying synthetic pesticides, using bees and other insects for pollination rather than human laborers, cultivating nitrogen-fixing plants rather than using synthetic nitrogen fertilizer, and recycling compost and animal waste generated on site rather than importing synthetic fertilizers.

Recognizing that one may design and build systems that share similar material traits to natural ecosystems allows one to respond to a criticism that has occasionally been leveled at biomimicry: that it involves substituting entirely artificial entities for natural ones, in which case, it is claimed, biomimicry runs the risk of leading to, and even legitimizing, the ever-increasing destruction of nature.[66] Examples that may appear to bear out this criticism include the use of artificial bees for pollination, artificial predators for pest control, and artificial trees for carbon capture. But this criticism overlooks the possibility that when imitating ecosystems we may imitate not only their forms but also their biotic and abiotic materials. A biomimetic system, such as a biomimetic city or a biomimetic farm, is necessarily artificial, for it is constructed and delimited by humans by arranging various materials in a certain form so as to carry out certain functions. But this does not mean that the materials out of which it is composed are also necessarily artificial, for it is quite possible to imitate the component materials of natural ecosystems, including their biotic materials, as is the case when natural bees are used for pollination.

A significant feature of the above analysis is once again the way it expands the scope of biomimicry. The majority of research that currently takes place under the rubric of biomimicry concerns the imitation either of forms or of composite materials (e.g., wood, nacre, spider's silk). But while the latter do indeed count as materials for another producer (e.g., someone making fabrics out of biomimetic spider silk), for their own producer they clearly count as forms that must be composed out of other more elementary materials. What typically gets overlooked, then, is the possibility of deliberately using the same materials as nature, from the simple atomic substances of which organic molecules are composed right up to the complex biotic elements—living beings, biological societies—of natural ecosystems. And to include this possibility allows biomimicry to cover much of what currently takes place in such fields as synthetic biology, the bioeconomy, and biodesign (designs which use living beings as components).

## Generative Processes

A third type of imitable trait that may be abstracted from nature are what I propose to call "generative processes." Generative processes correspond very approximately to what the Aristotelian tradition calls "efficient causes," though I have avoided this term on the grounds that it suggests that what are "caused" here are only effects, not beings. Aristotle's original discussion of this type of cause, however, includes both the bringing about of a change (effecting) and the bringing forth of an entity (producing). The term "generative process" is thus employed here on the grounds that it covers both of these processes; one may talk of both entities and effects as "generated." Nevertheless, it is important to remember that in the case of the generation of entities, the entities in question are not self-producing beings, but natural systems that only exist (as opposed to ensist) because they have been delimited and thus brought forth by humans.

Some examples of level 1 generative processes in living beings are the various different instances of biomineralization (the generation of minerals by living beings), the products of which often come to compose biological structures. So, just as one may imitate the many natural materials produced by biomineralization, including bone mineral in vertebrates, silicates in diatoms, and shells in molluscs, one may seek to imitate the generative processes of biomineralization whereby such materials are generated.[67] As for the generation of effects, as opposed to entities, level 1 examples in living beings would include the various processes by which organisms either increase or decrease their temperature, including sweating, modifying blood flow, and goose bumps, all of which could potentially furnish models for artificial mechanisms for temperature regulation.

Generative processes in living beings are also important sources of level 2 models for artificial production processes. A frequent observation made in biomimicry circles is that human industry currently employs a "heat, beat, and treat" approach: a succinct way of saying that our products are forced into shape using high temperatures, high pressures, and harmful chemicals.[68] Living beings, by contrast, generate new entities within their own bodies and so cannot have recourse to such violent production methods. We may thus seek to imitate such level 2 generative processes as making things within the normal temperature range favored by living beings (approx. 20 to 40°C); making things at relatively low, ambient pressures; and making things using chemicals not generally harmful to living beings, often using water as a solvent. Some of the key advantages of these approaches include reduced energy use, absence of toxic waste products, and ease of recycling.

Other important sources of level 2 models include the way that living beings often make things using an "additive" approach, which builds up structures and systems through growth, rather than subtracting them from a greater mass of material—a generative process that may reduce waste while making possible the realization of complicated biomimetic structures that could not be produced using traditional design and fabrication techniques.[69] This is not to say, however, that nature does not also have frequent recourse to "subtractive" approaches. Many of the cavities in the human body, for example, are initially produced by hollowing out, rather than by building around, and this in turn suggests the abstraction of more complex level 2 models in which additive and subtractive manufacturing techniques are artfully combined, with waste being either avoided or reduced through recycling off cuts.[70]

It is primarily with respect to generative processes that biological species provide important models for biomimicry. Evolution, understood as a process that generates favorable adaptations, is the obvious imitable trait here. An example is "directed evolution," which mimics the process of natural evolution in order to develop artificial substances (e.g., proteins, enzymes), and, in doing so, offers an alternative to the traditional practice of protein engineering by so-called "rational design."[71] As regards insect and animal societies, an interesting source of models here are not the constructions themselves (in their formal, material, and functional aspects), but rather the specific construction techniques employed, including site selection techniques, sequence of construction phases, modular growth strategies, and division of labor, all of which could potentially provide level 2 models for humans, most obviously in the fields of architecture, civil engineering, and urban planning and design.

Ecosystems are also important sources of generative processes. Regarding the generation of entities, an important model is ecological succession, understood as that process by which mature ecosystems are generated through a series of intermediary stages. Ecological succession provides the key model in analog forestry, but it could also be employed in farming, such that annuals (e.g., cereals, flowers) could be followed by perennial shrubs (e.g., berries), and later on by trees (e.g., fruit trees, softwoods, hardwoods).[72] Similarly, in restoration ecology, ecological succession may be adopted as a model such that, instead of directly trying to recreate the mature ecosystem initially destroyed or degraded, one initially seeks to prepare the way for, and perhaps also to guide or accelerate, the various stages of succession.

Ecosystems also provide important models for the generation of effects. A level 1 example of just such a generative process is evapotranspiration in trees, the cumulative effect of which is that forest ecosystems remain relatively cool even in hot weather. The generative process that produces this effect could potentially be reproduced in cities or other artificial systems, either by using the same biotic materials that produce the effect in nature (trees) or by mimicking the evapo-transpiration process itself, through designing buildings that pump water from the ground and release it into the atmosphere, just as trees do. A similar level 1 example, operative this time at the global level of Gaia, is the way $CO_2$ levels are kept low by the process of carbon sequestration that occurs in plants and soils. Again, this process could be imitated either by designing artificial systems in which plants and soils are important components or by mimicking natural processes of carbon sequestration.

An interesting level 2 model that may be abstracted from the above examples is the principle of "lots of littles." It is lots of little processes of evapotranspiration that cool the forest, and lots of little processes of carbon sequestration that cool the planet. The principle of "lots of littles" may in turn be contrasted with the principle of "one big" solution often found in human engineering. Rather than try to cool the planet by spraying particles into the upper atmosphere or dropping iron filings into the ocean, biomimetic approaches to geoengineering could adopt the level 2 model of lots of littles, including, for instance, lots of little tree planting initiatives carried out all over the planet. It is also interesting to note that a lots of littles approach is typical of how nature produces its various "goods" and may be contrasted with the "one big" approach often characteristic of contemporary human production. Most of the buildings we construct produce almost nothing (or at least nothing material), but some very big buildings, such as factories, generate huge amounts of goods, which are then consumed the world over. A lots of littles approach to production, by contrast, would make almost all buildings generative of goods (e.g., of energy via solar panels, of water via rainwater capture, of durable goods via 3D printing, of food via green roofs and walls)—a radical change with potentially far-reaching social, economic, and political implications.[73]

To conclude this section, let us briefly consider how the concept of generative processes encompasses various other traits sometimes included in typologies of imitable traits—namely, processes, mechanisms, and behaviors. The concept of generative processes is preferable to the bare concept of processes in that the imitable processes we abstract from nature are not just any old processes, but that subset of processes that, by virtue of either the entities or the effects they generate, may be taken as models for the artificial generation of entities or effects. As for the concept of mechanism, it may be understood as a specific subtype of generative process, one that generates an effect, but not an entity. It makes perfect sense, for example, to talk about sweating as a mechanism that cools humans, evapotranspiration as a mechanism that cools forests, and carbon sequestration as a mechanism that cools Gaia. But one would not speak of biomineralization as the mechanism, but rather as the process, by which bone is generated. As for behaviors, they also typically generate effects (even if some behavior involves also the generation of entities), but they differ from mere mechanisms in that they require the coordination of multiple mechanisms.

*Functions*

The fourth type of cause discussed by Aristotle are generally referred to as "final causes," which are what something exists in order to accomplish. If, however, we replace the notion of a final cause or telos with that of function, defined nonteleologically as "what something does," then the many problems associated with teleology in nature may be overcome. One may worry, however, that in this case there would not be four causes anymore, only three, with the fourth cause (the final cause) ultimately reducing to the third cause (the efficient cause). But there is a difference between *how* nature generates an effect (e.g., sweating) and the effect generated (e.g., keeping cool). And this difference is important, for it means one may imitate the effect generated, and therewith also the function, without imitating the process that generates it.

To this, it may perhaps be objected that, at least in the case of the living, the operation of natural selection means that forms, materials, and generative processes are adapted to function, in which case to imitate only the function, but none of the other traits thanks to which the function is carried out, would not be biomimetic at all. If it is true, for example, that nature "adapts form to function," then to imitate a natural function without imitating the form may even seem to go against nature.[74] But it is also true that, even in nature, form and function are not inseparable. The evolutionary concept of exaptation tells us that a form that originally had one function may acquire a different one in a different context.[75] Conversely, much the same function may be achieved in different ways in different beings, depending on the specific phylogenetic path taken. It follows that there is nothing inherently problematic or redundant about the idea that the scope of biomimicry may extend to the abstraction and imitation of natural functions.

In many cases, it is of course not just the function that one will seek to imitate. In the case of the Shinkansen or the self-cleaning surface of the lotus plant, what is imitated is both the form and the function. But the theoretical separation of function from other imitable traits is important, for it not only gives rise to the possibility of abstracting and imitating functions independently of other traits but it also expands the "design space" of biomimicry in such a way that it respects the important criterion of comprehensiveness.

There are many level 1 functions in living beings that one may plausibly seek to imitate, even without imitating any other basic traits. For example, the main function of leaves, which is to generate usable energy from the sun, may be imitated without imitating the generative process responsible for the function. Indeed, photovoltaic panels do not imitate photosynthesis, though they are often called upon by biomimetic designers to imitate the *function* of photosynthesis. When Michael Braungart and William McDonough take trees as models for buildings, they install photovoltaic solar panels on their buildings to imitate the solar energy-generating function of leaves, but without trying to imitate the generative process responsible for that function in nature.[76] Another example is provided by the darkling beetle, which has the ingenious ability to harvest water in the desert from fog. One way in which this function may be imitated is by imitating the microscopic structure of the beetle's forewings.[77] But it is also possible to imitate the function without imitating the traits that, in the case of the darkling beetle, allow the function to be carried out—specifically, by making nets that collect water droplets from fog.

At higher levels of abstraction, however, there is less scope for imitating the functions of organisms. As one proceeds to higher levels of functional abstraction, one arrives at such general functions as "rapid or efficient movement" or "conservation of energy," and these abstract functions are for the most part likely to be too general to provide useful models for human technology, if only because they are things we are already doing or trying to do. Nevertheless, it is perhaps worth noting that certain functions that today seem quite obvious, such as flying, may initially have arisen only through seeking to imitate nature. Were it not for birds and insects, the idea of constructing flying machines may never even have crossed our minds.

As for species, sexual reproduction provides a number of interesting models for functional imitation. One function of sexual reproduction is to mix the genes, and thus the phenotypic traits, of two individuals (the parents), when producing a third individual (the child). This function is present as well in the field of evolutionary computing, in which different solutions, known as "parents," mix with one another inside the so-called mating pool, thus giving rise to other solutions, known as "children." In evolutionary computing, various forms and generative processes present in sexual reproduction are also imitated, including the organization of information into "chromosomes" and the mixing of the information in "genetic recombination." But the function, arrived at through imitating nature, of creating new solutions through combining preexisting solutions, along with various other functions, such as finding globally optimal solutions (by natural selection) or enabling solutions to adapt to rapidly changing contexts, are nevertheless important models that may be abstracted from biological evolution and transposed over into the context of evolutionary computing.

In the case of biological societies, there is much scope for imitating the functions of the constructions of social insects. Passive temperature regulation, for example, is an important function of the nests of ants and termites and could potentially be achieved using quite different forms, materials, and generative processes from those of the social insects themselves. Of course, it could be objected here that many designers are already seeking to achieve passive temperature regulation and that, while it may help to turn to social insects for forms, materials, and generative processes that would enable this function to be achieved, little or nothing is gained by imitating only the function itself, for passive temperature regulation is something that human designers are already trying to achieve.

While there is indeed much that might be learned about passive temperature regulation techniques from studying the specific ways it is achieved in social insects, it is also striking to note just how prevalent passive temperature regulation techniques are. Social insects make almost exclusive use of these techniques, including nest positioning and orientation, ventilation systems, insulating materials, and collecting heat from the sun, with the only heat "actively" produced coming from the insects themselves (and associated micro-organisms) and without any equivalent to the active air-conditioning systems increasingly used by humans.[78] Thus it is that the *prevalence* of passive temperature regulation in social insects may underlie its adoption as a *general* principle in the field of human design, as opposed to one that is adopted only occasionally or sporadically, as is the case today.

Natural ecosystems are probably the most important source of functions for biomimetic design. Whereas in the case of biology many of the functions we might abstract from nature are already present in the field of human design, the same does not hold—or at least not until very recently—in the case of ecology. It is also important to note that there is a specific branch of ecology, functional ecology, concerned precisely with the different ecological functions played by biological entities, and that is of particular relevance to this area of biomimetic design. As in many of the cases discussed above, to imitate this or that ecological function, it may also prove necessary or appropriate to imitate various other traits thanks to which the function is carried out in nature. But there are possible exceptions. The function of rainwater infiltration, which in natural ecosystems is achieved by such things as tree roots and porous soil, may be achieved through quite different things, such as permeable roads and pavements.

Also of great importance in this context is the imitation of more abstract, level 2 or level 3, ecological functions. Such abstract functions as producing usable energy from the sun, enabling or effecting energy transfers between entities, recycling materials, creating niches for other organisms, adding redundancy into systems, and contributing to global climate regulation all provide important models for human design. As in the case of passive temperature regulation, many designers may already have adopted these functions, and not necessarily through imitating nature, but, once again, the prevalence of these functions in nature again suggests the importance of adopting them as general principles of human design.

## IMITATING NATURAL MODELS

### Mimesis in Art and Technology

The idea of developing a typology of imitable traits based on Aristotle's understanding of natural science as analysis of the four causes fits so well with his own thinking about technē as imitation of nature that it is in some respects surprising that he did not develop such a typology himself. But this lacuna is very much in keeping with the fact that, while the *Physics* is widely recognized as the locus classicus of Aristotle's view of technology as imitation of nature, it has relatively little to say about mimesis itself and thus about the process of transferring natural models to technological arifacts.[79]

While Aristotle did not discuss mimesis in any great detail in the context of technology, he did discuss it at length in the context of art and poetry, the subject of the *Poetics*. With this in mind, the present section will seek to develop a theory of mimesis based not on the rather limited discussion of it in the *Physics*, but rather on the more extended discussions in the *Poetics*.

Focusing on Aristotle's *Poetics* does, however, introduce two complications. The first complication is common in philosophical discussions of mimesis in art. Applied to art, in the sense of literature, music, painting, sculpture, and the like, but excluding technology and design, the word *mimesis* is almost always a quasi-synonym for representation and is nowadays sometimes even translated as such. As René Girard has pointed out, however, there are other contexts in which mimesis does not equate to representation.[80] Girard himself focuses on the relation between mimesis and desire. One often desires something because someone else desires it. But mimetic desire is clearly not a representation—in the sense of a portrayal or depiction—of the other person's desire.[81]

A similar point applies to the imitation of nature in technology. There is clearly a significant difference between, say, a painting or sculpture of a leaf and an attempt to produce a solar panel modeled on various traits abstracted from the leaf qua natural system. In the first instance, *mimesis* refers to an aesthetic representation—that is to say, a depiction or portrayal of something aesthetically pleasing or interesting, but of no direct practical purpose. In the second instance, *mimesis* refers rather to a functional reproduction, where functional reproduction is understood here as the reproduction of natural forms, materials, generative processes, or functions for directly practical purposes.

Importantly, however, the traditional tendency to pursue sustained philosophical inquiry into art, but not into technology, combined with the fact that philosophers only began to engage in sustained inquiry into technology *after* the traditional view of technē as imitation of nature had been explicitly rejected (by Hegel in particular), means that the overwhelming majority of philosophical thinking about mimesis has taken place in the context of art and not in the context of technology. The modern critique of technē as imitation of nature may have applied to art just as much as to technology, but the weight of the philosophical traditional nevertheless means that discussions of mimesis in the philosophy of art remain important to this day.[82] If, then, one is to bring Aristotle's discussion of mimesis in the *Poetics* to bear on the imitation of nature in technology, this discussion must be shorn of the traditional tendency to identify mimesis almost exclusively with aesthetic representation and must instead focus on mimesis in the traditionally neglected sense of functional reproduction.

The second complication is specific to Aristotle's *Poetics*. Mimesis is discussed here solely in the context of poetry, especially epic poetry and tragedy. This in turn engenders a number of important differences from mimesis in the context of technology: in poetry, the object of imitation is above all what Aristotle calls "men in action," rather than traits—corresponding, mutatis mutandis, to the four causes—abstracted from natural systems; the unity of an epic or tragic poem is provided by the plot (*muthos*), rather than by the main function of the artificial product or system; and epic or tragic poems elicit fear or pity on the part of an audience, rather than contentment or satisfaction on the part of a user or beneficiary. But, despite these substantial differences in context and application, at a more abstract and philosophical level, there is, as we will see in the following sections, a powerful conception of mimesis at work in the *Poetics* that, mutatis mutandis, may be productively transferred to the somewhat different context of biomimetics and bioinspiration.

### Transfer: Transposition, Translation, and Transformation

There are two basic processes involved in the imitation of models abstracted from nature: transfer and composition. Transfer, the focus of the present section, describes the process whereby a model abstracted from a natural system is in one way or another carried over to technology. Composition, the subject of the next section, describes the process whereby different traits are selected and combined together in the process of producing a given technology.

It is important to realize that the starting point of transfer are the models abstracted from natural systems, not the natural systems themselves. The models transferred, in other words, are forms, materials, generative processes, and functions—not entire natural systems, at least seen other than as complexes of these four traits. As for the process of transfer itself, in the classification system I propose it has three distinct modes: transposition, translation, and transformation. In the case of transposition, the model is exactly replicated. If, for example, the model is a form, then, if the form is exactly replicated such that the artificial entity is isomorphic with respect to the natural form on which it is based, that counts as an instance of transposition. In such a scenario, the other "causes" of the natural system from which the model form is abstracted may of course not be imitated at all or, if they are imitated, that imitation may involve other modes of transfer. But since transfer applies to imitable traits, not to the natural systems from which imitable traits are abstracted, the mimetic process in question—mimesis of form—would still count as an instance of transposition.

In the case of translation, there is an attempt to reproduce the natural model as closely as possible, and yet, because of contextual differences, certain changes are required. Imagine, for instance, that one intends to make an artificial entity based on the form of a natural entity, but, because the artificial entity is of greater size, a precise replication of its form would be suboptimal, and certain adjustments are thus required to ensure optimal functioning. It is not hard to see that there is an analogy here with linguistic translation. When one translates from one language to another, one typically aims to reproduce the meanings of the first language as closely as possible, but because the context is different—different words and expressions are available, literally equivalent words may have problematic connotations, grammars and sentence structures are not the same—one will be required to introduce a number of differences.

In the case of transformation, the third mode of transfer, there is no attempt to reproduce the natural model as closely as possible, given contextual constraints, but that is not to say that the natural model plays no role in the mimetic process. The natural model remains the point of departure for the artificial design, but any number of changes may be introduced, and not merely those required by the context. A simple example is a solar energy technology called "SmartFlower," the general form of which is clearly inspired by flowers, but that would not appear to transpose or translate the specific form of any actual flower.[83] Further, although I propose that one talk here of "transformation," this does not mean that the process in question concerns only forms. Materials, generative processes, and functions abstracted from nature may all be said to undergo processes of "transformation," provided that there is no attempt made to reproduce the model as closely as possible, given the contextual constraints.

With a view to understanding transformation more clearly, let us turn to Aristotle's *Poetics*. According to Aristotle, tragedy represents people as "better" than they are, and comedy as "worse" than they are. If we put aside the value judgments "better" or "worse," we arrive at the idea that imitations may be *substantially different from* the natural model on which they are based, and therefore also that the model may undergo a "transformation" when transferred over into the work of technē. One could perhaps object here that the only reason that people are depicted differently in tragedy and comedy from how they are in reality is because these genres create different contexts, and that is it these different contexts that are the source of the required changes. Tragedies and comedies, from this perspective, would not be transformations of the reality they represent, but only translations. There is, however, no obligation to adopt these genres; as Aristotle recognizes, we could equally well represent people "as they are," as he says is the case in the poems of Cleophon and the paintings of Dionysius.[84] It follows that what Aristotle is theorizing in his discussion of tragedy and comedy is not an attempt to represent people as closely as possible—that would call for realism rather than tragedy or comedy—but a creative transformation of the initial model, which, in the case of Aristotle's theory of tragic poetry, is undertaken for the sake of catharsis, but in the case of technology would be undertaken for other, more practical ends.

An analogy with a work of literature may shed light on this threefold understanding of transfer. When Homer's *Iliad* is reprinted in ancient Greek for purposes of study by classical scholars, that is an instance of transposition. When it appears in modern English, that is obviously an instance of translation. And when it is taken as a source of inspiration for a quite different work of art, as in the case of William Shakespeare's *Troilus and Cressida*, that is an instance of transformation. In order for the analogy to be exact, however, one would not consider the two works directly (the original, on the one hand, and the reprint, translation, or new work inspired by the original, on the other), but rather the transfer of various imitable traits abstracted from the original, some of which might be simply transposed, others translated, and yet others transformed. Thus it is that a work of art may employ the same materials as Homer's *Iliad*, or at least some of them (there are a great many "materials" in a work as complex as the *Iliad*), while transforming the form (or at least some of the forms). Shakespeare's *Troilus and Cressida*, for example, features many of the same characters—understood here as materials—as the *Iliad*, but, through the use of poetic licence, famously deviates from the tragic form of Homer's original by introducing elements of comedy. When extended into the realm of technology, one may thus say that the core of transformation is not quite *poetic* licence, but rather *poietic* licence, understood as intentionally deviating from or paying only relatively scant attention to the natural model, and for reasons other than just contextual constraint, in the production of the technological entity.

Another important issue to consider here is whether translation and transformation may occur with respect to natural models operating at higher levels of abstraction (e.g., levels 2, 3). In theory, this is certainly possible, and yet there is, generally speaking, less scope for translation, and especially transformation, at these higher levels. The reason for this is not hard to discern. In abstracting from, say, level 1 to level 2, certain features present only at level 1 are lost, such that these features can no longer undergo transformation at this higher level of abstraction. In keeping with this, it is also important to note that the *translation* or *transformation* of a level 1 trait will often involve the *transposition* of a level 2 trait, for, even if the details of the trait being abstracted are modified its more abstract features may well be preserved. The same results may potentially be achieved, then, either by abstracting a level 2 model and transposing it over to the artificial entity, or by abstracting a level 1 model and then translating or transforming it in the production of the artificial entity.

This is not to say, however, that the two processes are methodologically equivalent. In the former instance, details of the model are lost, such that one has relatively little guidance as to how one may realize an artificial design based on it. In the second, the details of the model are not lost, but, on the contrary, undergo translation or transformation during the production of the artificial entity. It is one thing, for example, to abstract from the leaves of the lotus plant the level 2 model of a microscopic surface structure (form) that cleans itself (function) and then seek to make artificial entities that also possess these higher-level forms and functions, and quite another thing to abstract detailed level 1 models of the structure and function of the surface of the lotus leaf itself, and then to introduce differences, such as protrusions whose form make them less likely to break off, as may be required by the different materials used or by the inability of the artificial structure to renew or heal itself (an instance of translation).

An important consequence of the above analysis of the three different modes of transfer is that it allows us precisely to determine the difference between biomimetics and bioinspiration. Discussing biomimetics, Julien Vincent and his colleagues make the following observation: "A simple and direct replica of the biological prototype is rarely successful, even if it is possible with current technology. Some form or procedure of *interpretation* or *translation* from biology to technology is required."[85] Biomimetics, it would seem, involves either transposition or more often—since exact replication is "rarely successful"—translation. Bioinspiration, by contrast, does not introduce modifications simply because they are required in order for the functional reproduction to be viable, but, on the contrary, transforms the initial model for any number of other reasons.[86] Moreover, the fact that a lower-level transformation (i.e., an instance of bioinspiration) may count as a transposition or translation (i.e., an instance of biomimetics) at a higher level of abstraction provides insight into why some researchers may prefer the term "bioinspiration": they seek to imitate nature at higher levels of abstraction, especially at the level of "underlying principles," rather than to transpose or even translate specific level 1 models.

### The Art of Composition

The other important process involved in imitation, composition, is the one discussed in most detail by Aristotle. Consider again Homer's *Iliad*. On an Aristotelian reading, the *Iliad* is not, as it would have been for Plato, an inevitable failure to represent the people and events of the Trojan War, themselves degraded instances of certain Ideas (e.g., honor, justice, love), but a new composition not found in nature. Some of this difference from nature may of course be accounted for by translation and transformation—that is to say, by deliberately introducing changes with respect to the initial model, especially representing people as better (e.g., stronger, braver) than they are in reality. But other differences concern composition: the selection and combination of various traits abstracted from nature so as to form a composition not present in nature. Homer does not, for example, attempt the impossible task of depicting every last detail of the Trojan War, including insignificant participants and banal conversations, but instead selects and combines various different elements on the basis, Aristotle would say, of their relevance to the plot.

Despite the quite different context, the notion of composition may also be applied to contemporary biomimicry. To produce an artifact, Aristotle tells us, four different "causes" must be brought together, with the art of composition lying in skilfully selecting and combining these—a mode of legein specific to technē. The Shinkansen, for example, imitates the form of the kingfisher's beak, but not its materials or generative processes. As for the beak's function, it is only the relatively abstract, level 2 trait of smoothly passing from one physical context to another that is imitated, for other functional features of the beak, such as passing into water or capturing prey, are manifestly ignored. The composition of the Shinkansen, it follows, involves abstracting a level 1 form and a level 2 function from nature and combining these with materials and generative processes not based on natural models. But that is of course only one example of composition. In theory, any combination of natural and artificial "causes" could be combined and articulated in the production of a biomimetic artifact, provided only that at least one of these "causes" come from nature (otherwise the artifact would not be biomimetic at all, but conventional). Even making an object out of wood, but through imposition of an artificial form, using artificial generative processes, and for artificial ends, could be undertaken biomimetically with respect to the material cause. In most cases, however, multiple traits will be abstracted from the natural system in question and combined together in the process of biomimetic design.

Another important possibility is what one might call "biomimetic hybridization," whereby traits abstracted from quite separate natural beings are selected and combined together. An example is a recently invented "dual biomimetic surface" that combines the surface morphology of shark skin with the hierarchical micro/nanostructures of the lotus leaf.[87] Similarly, one could also imagine cases where a new trait is produced by selecting and combining instances of the same trait abstracted from two quite different beings. One could, for example, produce the aerodynamic form of the wing of a plane by selecting and combining two similar forms (e.g., the forms of two wings) abstracted from quite different animals (e.g., two species of birds). Such nature-nature hybrids may be contrasted with what is at present the more common case of nature-artifice hybrids, which combine natural and artificial "causes" (e.g., natural forms with conventional man-made materials), as in the example of the Shinkansen.

A final important case to consider here are "complex" artifacts. In Aristotle's examples, the artifacts in question—silver chalices, wooden beds, bronze statues—are "simple," by which I mean that they combine one form, one material, one generative process, and one function. But there are also "complex" artifacts that combine multiple different forms, materials, generative processes, and functions. Consider an aeroplane: its forms include not only the external structure (e.g., wings, body, tail) but also the internal chassis, the shape of the seats, the layout of the cabins, the complex circuitry of the on-board electronics, and the mechanical structure of the engine. Much the same principle obviously applies regarding the plane's materials, generative processes, and functions. So, while there may be one basic natural system from which models are abstracted—namely, the winged bodies of birds, for it is this that initially provided planes with the models for their basic form and function, albeit primarily at higher levels of abstraction (e.g., general principles of winged flight, aerial travel from one location to another), there is clearly plenty of scope in the case of an aeroplane for drawing on other natural models, many of which could potentially be abstracted from quite different natural beings.[88]

### Creative Imitation

On the basis of the above analyses of transfer and composition, it is clear that drawing on Aristotle gives rise to a positive view of mimesis which, as such, is opposed to the negative view put forward by Plato. For Plato, there is an inevitable loss of form involved in imitating nature. Just as a photocopy aims for faithful reproduction of an original document, but never quite succeeds, the same is true of works of technē, in which case one may perhaps speak here of "destructive imitation." For Aristotle, by contrast, mimesis is seen in a positive light as a creative process that gives rise to new forms and compositions not present in nature. This, then, is why Paul Ricoeur attributes to Aristotle a theory of what he calls "creative imitation."[89]

The concept of creative imitation makes it possible to overcome both the traditional view of technē as faithful imitation and the more recent view of technē as pure creation. Creative imitation overcomes the lack of creativity involved in faithful imitation by allowing for two main ways in which we may introduce modifications with respect to natural models: transformation and composition. At the same time, nature still provides models for the design of products and systems that—in contrast to the ideal of "pure construction and synthesis" characteristic of modern technology[90]—no longer see the rejection of natural models as a virtue, as if taking nature as model were an affront to human creativity.[91]

If creative imitation avoids the extremes of both faithful imitation and pure creation, then it follows that what one might call "technical virtue" may lie between these two extremes. Just as, for Aristotle, ethical or practical virtue lies between two extremes (e.g., courage as lying between cowardice and recklessness), technical virtue may lie between the two extremes of faithful imitation and pure creation. Anyone who goes out of their way to achieve exact imitation of a natural model (transposition), or even just to limit the mimetic process to the attempt to imitate the natural model as faithfully as possible given contextual constraints (translation), is not only drastically limiting their creative freedom but also reducing the likelihood of producing a successful technological object. However, the converse is also true. Someone whose approach is limited to trying to design new technologies without any natural models to guide them will also be artificially restricting their creative potential. By contrast, someone who is sufficiently knowledgeable and resourceful to be aware of a variety of relevant natural models, to understand how they work, and to be able to select, transfer, and combine them so as to form new products will find their creativity enhanced. The relation between imitation and creation, we may conclude, is one of complementarity, rather than of antagonism or opposition.

## BIOMIMETIC INNOVATION SCENARIOS

### Biological Push and Technological Pull

In the previous sections, I presented and explored the various possibilities involved in the process of biomimetic innovation, which proceeds by selecting from a vast set of possible models at varying levels of abstraction and then imitating them via various different modes of transfer and composition. In this and the following two sections, I consider how, within this huge expanse of possibilities, various different types of innovation scenario may arise, where by an innovation scenario I mean a specific way in which the general process of abstraction, transfer, and composition common to all biomimetic innovation may be realized in practice.

The first two innovation scenarios I shall discuss are usually referred to in the academic literature as "biological push" and "technological pull."[92] In cases of biological push, one starts with knowledge of one or more traits abstracted from nature and then either looks for or immediately sees a technological application. This may arise through discovering some new trait hitherto unknown to science, through better understanding a trait already known but inadequately understood, or through realizing that a trait that is already well understood may have a hitherto unnoticed biomimetic application.

A classic example of biological push is the invention of Velcro by the Swiss engineer George de Mestral. While walking his dog in the Alps, de Mestral noticed that the burrs of the burdock thistle had stuck to his pet's fur. He became curious as to how this worked and, upon inspection, discovered hundreds of tiny hooks that caught on to anything appropriately loop-shaped, the hairs of his dog included. This in turn provided the occasion for his idea of imitating the form and function of the burdock thistle's hook and loop fastener, though not its materials (originally cotton, later nylon), or, as far as I know, any of the generative processes involved.

The opposite of biological push is technological pull. In technological pull, one starts with a technological problem—a problem of how to design something that carries out a desired function—and then looks to nature for a solution, or at least for certain elements of a solution. A classic example is the invention of the Shinkansen, which began with the problem of how to make a train that would pass regularly in and out of tunnels and that was insoluble using conventional forms, for these created noise and turbulence. This problem then prompted a turn to nature for help finding a solution and, more specifically, to the form of the kingfisher's beak.

Many other cases of technological pull lead to solutions drawn from ecology. Indeed, it could perhaps be argued that it is a general crisis in the nature of modern technology (its unsustainability, its inappropriateness) that is currently driving the contemporary turn toward ecological models in the technical disciplines (e.g., engineering, architecture, design). This turn may then lead to a general recognition of nature's "genius," and in particular of its general suitability as a source of models for effective, appropriate, and sustainable technologies. But the initial motivating factor for the turn to nature is in many cases a recognition of the radical ecological shortcomings of existing technologies, rather than, as is the case in biological push, recognition of nature's genius and a desire to imitate it.

## Biomimetic Analogy and Simile

Any biomimetic technology will in some respects be analogous or similar to its natural counterpart, for if that is not the case then there is no meaningful sense in which it could be said to imitate it. But how are we to understand the notions of "analogy" and "similarity" here? By definition, the notion of "analogy" implies that there is some sort of parallel or correspondence between the *logoi* of two entities or systems. As we have already seen, the root meaning of the word *logos* is "gathering" (legein), and what I suggest is gathered together in the natural and technological systems under discussion are precisely the four causes identified by Aristotle. In view of this, to say that there is an "analogy" between a natural and an artificial system is to say that there is a parallel not simply between this or that trait (e.g., both having more or less the same form, or both being made of more or less the same materials), but rather between the way their traits are selectively brought together. By contrast, in cases where there are parallels between only one trait, or where there are parallels between multiple traits but not between the way they have been selectively combined, one would not speak of analogy, but only of similarity.

Analogies and similarities between natural and technological entities are not, however, only *consequences* of biomimetic innovation scenarios, such as biological push or technological pull. It is also possible for a biomimetic innovation scenario to *begin with* the observation of an existing analogy or similarity between a natural entity and an artifact. This analogy or similarity may have arisen by previous deliberate imitation, by having been generated in comparable contexts (as happens also in the phenomenon of convergent evolution studied by biologists), or, at least in the case of similarity, simply by chance. But this starting point in analogy or similarity only gives rise to biomimetic innovation when it leads to the exploration of the ways in which the artificial entity could be made *more similar or more analogous* to the natural one.

Noticing that cameras and human eyes are analogous to each other with respect to form and function could provide the basis for further innovations that would make cameras *more like* human eyes—for example, by using similar mechanisms to adjust quickly and effectively to low levels of light; by being connected up to information processing systems capable of identifying various types of entities (albeit without, in the case of cameras, actually understanding them as such); by having shutters that, like human eyelids, wipe their lens clean of dust; by having two lenses capable of generating stereo pictures (as already exists in stereo photography); by being able to self-heal when the lens is scratched; and by having spherical lenses capable of rotating in artificial sockets.[93]

Another example, which this time gives its analogical status away in its very name, is the case of analog forestry already mentioned above. Taking as its starting point the obvious similarities between natural forest ecosystems and tree plantations, the basic aim of analog forestry is not just to make the latter more like the former, but also, and more profoundly, to move from similarity to analogy, primarily by imitating the natural *logic* (in the sense of logos) of ecological succession: the way the various processes involved in ecological succession generate various formal, material, and functional traits and, in so doing, provide forestry with a powerful analogical model.

It is instructive to compare analogy and simile in biomimicry with analogy and simile in language. One of the functions of linguistic analogies and similes is of course simply to show that two things are analogous or similar in some relevant respect. But this recognition often leads one to try to pursue the analogy or simile, thus making the things appear even more similar or analogous than they did before—a process that often involves developing an initial simile into a bona fide analogy. Much the same is true of biomimetic analogies and similes, though in these instances it is not only in our minds that the already analogous or similar things become more analogous or similar (though this becoming more analogous or similar will usually occur first in our minds) but also in the realization of the biomimetic design. As in the case of linguistic analogy and simile, however, there is a danger of overstretching the comparison. To design cameras with eyebrows and eyelashes may perhaps prove a step too far.

It is also important to realize that, just as linguistic analogies and similes are typically drawn in order to shed light on only one of the two entities involved, in cases of biomimetic analogy and simile it is one entity (the natural model) that is called upon to shed light on another (the artificial imitation). And yet, as in the case of linguistic analogy, it is always possible that the analogy or simile will end up having the reverse function, that an analogy or simile drawn between A and B to shed light on B may end up shedding light on A. Applied to biomimetic analogies and similes, this implies that analogies or similes between nature and technology may also in some cases shed light on nature, such that in making our technologies more analogous or similar to nature we may also obtain a better understanding of nature—an idea reminiscent of Kapp's view that it is through understanding the artifacts that unconsciously replicate the forms of the human body that we obtain self-understanding.[94] We may, for example, gain insight into how natural brains work by making artificial ones.

There is, however, an important epistemological limitation here. What we may come to understand in nature by analogy or simile with biomimetic technologies could extend only as far as the logic underlying the gathering together of forms, materials, generative processes, or functions in a natural system, and never to nature itself in the sense of physis. For there is of course a fundamental *dis-analogy* between technē and physis that no technology, no matter how sophisticated, could ever overcome: whereas physei-beings are self-producing, technē-beings are produced by something other. And if it is true, as Giambattista Vico famously argued, that we can only ever fully understand what we ourselves can make (*verum esse ipsum factum*), this implies that we could never fully understand natural beings, for, as self-producing, they are by definition not producible and—ex hypothesi—fully understandable by us. It is perhaps at least partly in this sense, then, that we must understand Heraclitus's idea, taken up in recent times by Heidegger, that what produces and presents itself at the same time also conceals itself (*physis kryptesthai philei*).[95] As that which can never be produced by us, physis would also be that which we can never fully understand.

## Biomimetic Metaphor

Innovation by biomimetic analogy or simile differs from the better documented scenarios of innovation by biological push and technological pull in that it takes as its starting point not one thing, either a natural trait (biological push) or a technological problem (technological pull), but two things, between which it establishes some sort of comparison. Biomimetic metaphor shares with biomimetic analogy and simile a starting point in two existing things. And yet, as will soon become apparent, it is also quite different. To understand this difference, I will first turn my attention to developing a philosophical theory of metaphor, before going on to transfer that theory into the specific context of biomimicry.

Aristotle shall once again provide the point of departure for the present discussion of metaphor. As the first philosopher to have paid explicit attention to metaphor, he also set the stage for future discussions of the topic. This is not to say, however, that I will seek to return to and renew Aristotle's theory of metaphor; it is precisely this theory that, I think, must be overturned if we are to develop a viable and significant theory of biomimetic metaphor.

In the *Rhetoric*, Aristotle claims that metaphor and simile are basically the same, the only difference being the form of expression, and this is indeed how the two are usually viewed today.[96] Metaphor, from this perspective, is simply a condensed or elliptical comparison; it draws attention to analogies or similarities between two things, and its raison d'être lies solely in being more aesthetically satisfying or economical. Max Black calls this the "comparison view" of metaphor.[97] Black notes, however, that there is another traditional view of metaphor, which he calls the "substitution view."[98] According to this latter view, when speaking metaphorically one simply substitutes a figurative description for a literal one. From this perspective, metaphors provide no information content that could not have been provided by a literal description and their raison d'être is only to provide aesthetic decoration or to allow one to say the same thing more economically than by speaking literally. What these views share in common is the belief that metaphor is reducible, at least as far as what Black calls its "cognitive content," to something else, whether a comparison or a literal description.

Black himself puts forward a third view of metaphor, which he calls the "interaction view." As its name suggests, this view holds that metaphors bring two entities into interaction with one another, such that the meanings, connotations, and implications associated with one entity may be transferred or carried over (*metapherein* in Greek) to the other. It is not immediately obvious, however, how exactly these interactions differ from any that occur in analogy or simile. Black at one point suggests an answer to this question: metaphor is at least in some instances creative, for, whereas comparisons involve only *perceiving* similarities between things, metaphors in some instances *create* similarities between things.[99] From this perspective, what metaphors make similar are not only our understandings of the things concerned but also the things themselves. But, as Black's critics have pointed out, to say that metaphor creates similarities between things, as opposed to simply pointing out antecedent similarities as occurs in comparison, is clearly open to the common-sense objection that a mere figure of speech has no such power to change the things themselves, in this instance by creating similarities between them.[100]

If metaphor is not different from comparison by virtue of its capacity to create similarities between things, how is it different from comparison? The answer, I suggest, is that metaphor does not, as is the case in comparison, invite us simply to *perceive* similarities between things; rather, it invites us to *imagine* similarities between things. What, from this perspective, takes place in metaphor is neither a pointing out of the similar, nor the creation of real similarities between the antecedently dissimilar, but an act of the imagination whereby we counterfactually entertain the similarity of the dissimilar. Between the first extreme of reducing metaphor to comparison, so that it becomes simply an elliptical or condensed way of drawing attention to the antecedently similar, and the second extreme proposed by Black, whereby metaphor is endowed with a creative power to change the things themselves by making them similar to each other, lies a third possibility: that metaphor involves imagining that the dissimilar is in fact similar, but without actually making it so. The key to metaphor, it follows, lies not in our powers of perception or creation, but in the power of the imagination. Let us call this the "imaginative view" of metaphor.

The imaginative view explains why metaphor so often makes great interpretive demands on its audience. Whereas analogy and simile lay out existing similarities between things, metaphor asks us to use our imagination to conceive similarities between things and, in so doing, makes possible the transfer of various traits from one entity to the other. So, whereas the starting point of analogy is "what is" (i.e., what is similar), the starting point of metaphor is rather "what if?" (i.e., what if we imagined two dissimilar things as similar?). And, in keeping with the general call that "what if?" questions make on our imagination, there is a potentially unlimited reserve of associated traits that may be transferred across from one thing to another when, by means of metaphor, we are called upon to imagine that thing as like another quite dissimilar thing. This in turn allows us to identify a major problem in the substitution view of metaphor. In metaphor, there is no clearly defined set of literal meanings that the metaphor economically presents in a single word and that may be uncovered by interpretation; on the contrary, the traits that metaphors make it possible to transfer from one entity to another are not clearly delimited or circumscribed in advance, but instead emerge only through imaginative interpretation.[101]

Before going on to consider how this imaginative view of metaphor may be transferred into the context of biomimicry, let us consider one obvious and important objection: that the imaginative view does not well fit all cases or instances of metaphor. Indeed, it seems obvious to say that in many cases metaphor does not take as its starting point two dissimilar things and then ask us to imagine them as similar, but, on the contrary, takes two similar things and merely reveals or underlines their antecedent similarity, as in the case of the comparison view. In short, there would appear to be many instances in which metaphor only differs from comparison by virtue of its linguistic form (the metaphor of A *as* B, rather than the description of A as *like* or *analogous* to B).

There are two ways of replying to this objection. The first is to note that it takes the linguistic form generally held to be characteristic of metaphor as the standard against which a philosophical theory of metaphor is judged, with the almost inevitable result being that the philosophical theory never quite measures up. There is, however, another way of looking at this issue: by seeing the imaginative view of metaphor as the true one and apparent instances of metaphor that do not fit that view not as counterexamples or anomalies, but simply as inappropriate ways of expressing literal descriptions or comparisons. From this perspective, the general correspondence between the textbook form of metaphor (the metaphor of A as B) and the nature of metaphor, as revealed by philosophical inquiry, arises only because of the general appropriateness of the former for representing the latter. And yet it is also possible, firstly, that the textbook form of metaphor may be employed to express something quite different from the true nature of metaphor (such as literal description or comparison), in which case the use of the textbook form of metaphor is ultimately misguided, and, secondly, that true metaphors may be expressed in inappropriate forms, such as comparisons or even apparent attempts to speak quite literally, as occurs when people speak of things "literally" occurring that only make sense as metaphors (e.g., "It was literally raining cats and dogs").

The second way of replying to the objection in question consists in arguing that many instances in which metaphor appears to be but a kind of elliptical or condensed comparison only appear this way to us because the metaphor is so familiar. Recall that the nature of metaphor is to make one imagine the dissimilar as similar. Now, when we are called upon to do so again and again with respect to the same pair of entities, we may easily come to believe that the dissimilar is in fact similar, and not simply that we are only imagining it to be so. When love is described as a red rose, or sadness as winter, we are so used to these metaphors that we may easily come to believe that this is because the things they describe are in fact similar—that there is, to pursue the above examples, some sort of real similarity between love and roses or between sadness and winter, rather than just familiar metaphorical associations between the two. Metaphorical familiarity is easily mistaken for real similarity.

In making this stark distinction between the similar (the analogous included) and the dissimilar, the question is of course raised as to how it is that the two may be distinguished, such that we might in every case be able correctly to identify candidate cases of true metaphor (or true comparison). In response to this question, one may draw on the concept of *das Man* set out in the previous chapter, and thus also of "understanding-as-one-understands." What this concept allows us to see is that, while there is no absolute and objective way of categorizing things as similar or dissimilar, there is nevertheless an average, everyday understanding of whether different things are similar or dissimilar. Metaphor, from this perspective, does not invite us to imagine two things that are objectively dissimilar as being objectively similar; rather, it invites us to imagine as similar two things that "one" (*das Man*) does not habitually understand as similar (outside the specific context of metaphors). Metaphor, in other words, is relative to our average, everyday understanding of what is similar and dissimilar, not to absolute and objective reality.

Let us now turn our attention to biomimetic metaphor, understood here as an innovation scenario. In light of the imaginative view of metaphor, innovation by biomimetic metaphor may be said to differ from innovation by biomimetic comparison in that it takes as its starting point not two already similar entities, but two dissimilar entities, one natural, the other artificial, and then calls on us to imagine the artificial entity as, or as like, the natural entity. At the same time, biomimetic metaphor retains an important feature of biomimetic simile and analogy. Just as biomimetic simile or analogy differs from linguistic simile or analogy in that it does not just draw a comparison between two entities, but instead takes the initial comparison as a starting point for making the two entities more similar or analogous in reality, so biomimetic metaphor does not simply imagine the dissimilar as similar, for it subsequently proceeds to make the dissimilar similar in reality, thus realizing what was previously only imagined.

As a case study of biomimetic metaphor, consider the following saying of Braungart and McDonough: "Imagine a building like a tree, a city like a forest."[102] It is not hard to see that this expresses well the imaginative view of metaphor: despite the presence of the word "like," which would seem to imply simile, what we are called on to do here is not to perceive similarities, but, on the contrary, to imagine similarities that were not there before. To say "imagine a building like a tree," is thus equivalent, at least in terms of its cognitive content, to proposing the biomimetic metaphor of "buildings as trees," for in both cases what we are invited to do is to imagine the dissimilar as similar.[103] And, indeed, there can be little doubt that trees and buildings, cities and forests, are not similar entities, or, more precisely, that "one" would not generally consider them as such. Terraced houses, for example, adjoined as they are to other houses, approximately square in shape, with pitched roofs, hollow rectangular interior spaces, drainpipes, and windows have next to nothing in common with the forms discernible in, say, an oak tree. And much the same will generally be true also of the materials, generative processes, and functions of these houses.

Of course, one could perhaps object that, in certain cases, or at sufficient levels of abstraction, similarities may start to become visible. Detached houses may be said to resemble trees in that they are relatively discrete entities, sticking out from the ground, with spaces between them, and of approximately comparable scale. But even if it is true that if one looks hard enough similarities may well be discerned, in the case of the exercise of the imagination proposed by Braungart and McDonough, these similarities are manifestly not the starting point of the innovation scenario. On the contrary, the starting point is precisely the radical *dissimilarity* between buildings and trees, cities and forests, which the proposed exercise of the imagination invites us to overcome. Here, then, we have an innovation scenario that is clearly quite different from those cases in which it is the perception of an analogy or similarity between two entities that provides the occasion for their becoming *more similar*. Just as metaphor, on the imaginative view, cannot be reduced to analogy or simile, the same applies also to biomimetic metaphor.

It is also instructive to consider the failings of the substitution view of metaphor, transferred here into the context of biomimicry. Just as the substitution view holds that the function of metaphor is purely decorative or economical, in which case one could ultimately say exactly the same thing quite literally, to transfer the substitution view of metaphor into the context of biomimicry implies that biomimetic metaphors are but decorative or economic ways of describing designs that could easily have been arrived at and spelt out without the biomimetic metaphor in question. From this perspective, saying that buildings should be imagined like trees is just an economical or decorative way of saying that they should be designed in familiar "green" or "environmentally friendly" ways, such as running on solar energy or being made of recyclable materials. The main problem with this is that it fails to recognize the various ways in which biomimetic metaphor may enable us to imagine artifacts that are quite different from anything we might otherwise have imagined. To imagine buildings like trees provides an occasion to imagine a building that possesses all sorts of traits that go beyond mainstream attempts to make buildings more environmentally friendly, including the ability to transfer excess energy to a distributed underground energy network (modeled on the rhizosphere), to produce edible leaves and fruit, to cool its surroundings by evapotranspiration and shading, to facilitate rainwater infiltration, to enrich the soil, to provide a habitat for local fauna, to offer physical support for creeping plants, and to change with the seasons.[104]

The example of "buildings as trees" highlights another important feature of biomimetic metaphor. As a general rule, in the case of biomimetic metaphor the ultimate source of models is a complex entity, such as a living being or an ecosystem, which is not taken here as a self-producing whole, but rather as a complex system, or a complex ensemble of subsystems, from which imitable traits may be abstracted.[105] A tree, from this perspective, is not a self-producing being, for, as such, it cannot be imitated, and yet neither is it reducible simply to one single trait (e.g., carbon sequestration or solar energy generation); rather, it is a collection of natural systems (e.g., leaves, xylem, roots) in interaction with one another and with the tree's external environment, and from which many traits may be abstracted and transferred over to the design of buildings. This, in turn, allows us to see an important difference between, on the one hand, biomimetic metaphor, and, on the other, biological push and technological pull: in the latter cases it is typically one specific trait or ensemble of traits (or the need thereof, in the case of technological pull) that lies at the origin of the innovation scenario. The Shinkansen, for example, was developed not by imagining a train like a kingfisher—that is to say, via the biomimetic metaphor of the "train as kingfisher"— but by identifying a specific problem in existing trains and then abstracting the relevant formal and functional trait from an appropriate natural system: the kingfisher's beak.

It is also important in this context to take note of Black's observation that the explication of metaphor, the actual unfolding of its various implications, involves a certain loss of "cognitive content."[106] Transferred into the context of biomimicry, this means that there is a wealth of possible implications in a biomimetic metaphor that any corresponding design will inevitably fail to realize in full. The biomimetic metaphor of "buildings as trees, cities as forests" has an almost limitless range of potential implications and ramifications, such that different designers would likely take it as a starting point for the realization of significantly different artificial products and systems, depending on the specific processes of abstraction, transfer, and composition they choose to carry out. Any particular building or city, while unfolding some of the possible implications of this highly fertile metaphor, will inevitably imagine and realize only some possible implications, leaving others concealed and unrealized.

Another important issue to consider is the transfer of various different types of metaphor, including extended metaphors, mixed metaphors, and dead metaphors, into the context of biomimetic innovation. In literary criticism, an extended metaphor is a metaphor to which the author returns repeatedly and further develops over the course of their work. Transferred into the context of biomimicry, an extended biomimetic metaphor would involve the repeated return to and explicit development of imagined similarities between initially dissimilar natural and artificial entities. An architect who returned again and again to models abstracted from trees when designing a building could thus be said to have taken trees as an extended metaphor for buildings. Similarly, it is possible that the extension may concern the wider contexts in which the first use of the metaphor was embedded, as occurs when Braungart and McDonough extend the biomimetic metaphor of buildings as trees to cities as forests. Extended metaphors are only really possible, however, in the case of complex natural beings or systems, from which it is possible to abstract a significant number of different imitable traits, as is the case regarding trees and forests.

Mixed metaphors provide another interesting case. Unlike extended metaphors, which involve returning to and developing the same metaphor, mixed metaphors involve the mixing of different metaphors. Traditionally, mixed metaphors have been prohibited or at least deemed problematic, generally on the grounds that they are contradictory or confusing. This same flaw may also surface in biomimicry. Consider the following hypothetical instance of a mixed biomimetic metaphor: "Imagine a building like a tree, a city like a polar bear." It is hard to see how one might make sense of this. Polar bears are not made of trees, so if a city were made to a large extent of buildings like trees, it is hard to see how, as a whole, it could possibly come to resemble a polar bear.

And yet, just as an outright prohibition of mixed metaphors would be unnecessarily restrictive, for it is quite possible that, via mixed metaphor, one thing may without confusion or contradiction be imagined as similar to multiple other things—namely, by imagining it as similar to one thing in some respects and to another in other respects—so it is quite possible to formulate fertile mixed metaphors in biomimicry by imagining an artificial entity as similar to one natural entity in some respects and to another natural entity in other respects. For example, one could make sense of the idea of imagining a building that was like both a tree and a polar bear by imagining that, in addition to the transfer from trees to buildings of the sort of arboreal traits presented above, other traits present in polar bears, such as having transparent hair that allows the skin to absorb sunlight, using highly effective insulation materials, adapting blood circulation to external temperatures, or having a high volume to surface ratio, could also be abstracted and transferred over to buildings.[107] In doing so, one would of course have to be wary of not creating contradictions, incompatibilities, or even just suboptimal trade-offs, but there is no reason to exclude mixed biomimetic metaphors a priori. Indeed, mixed biomimetic metaphors are but particular instances of what I earlier called "nature-nature hybrids" (two or more natural models sourced from different beings) and the art of mixed biomimetic metaphor but a specific instance of the art of biomimetic composition.

Another important type of metaphor to consider here are dead metaphors—that is, metaphors that are no longer recognized as such. Dead metaphors cannot lie at the origin of biomimetic innovation scenarios. Biomimetic innovation always involves intentionally or deliberately taking a natural entity as a model for some sort of artifact. Of course, an artifact may still be biomimetic even when its designer or producer is not aware that the preexisting design on which their new design is based derived from a natural model (or set of natural models). Someone who imitated a building based on the model of a tree, without realizing it was based on a tree, would still have made a biomimetic artifact. But in such instances only the artifact is biomimetic, not the innovation process. And it is precisely in instances such as these, when the original biomimetic metaphor is no longer active, that one may talk of "dead metaphors." Dead metaphors in biomimicry, it follows, are but special instances of what one might call "dead models"—that is to say, models that have been abstracted and transferred over to technologies—including through other biomimetic innovation scenarios—but that have since been forgotten.

An example of a dead biomimetic metaphor may be found in organic farm-ing. Organic farming was originally developed by Albert Howard, who developed the insightful biomimetic metaphor of the farm as forest.[108] Many people prac-ticing organic farming today, however, are completely unaware of its origin in bio-mimetic metaphor, in which case the metaphor may be said to be a dead one; organic farmers still reproduce some of the implications of that metaphor every time they farm organically, but for the most part they do so unknowingly. A more problematic example concerns anthropomimicry. In a historically important instance of anthropomimetic metaphor, urban designers from the Renaissance took human beings as models for cities, including, from the nineteenth century, with respect to their internal forms and functions.[109] To the extent that contem-porary urban designers, engineers, and developers unknowingly reproduce these designs, they may also be said to be unknowingly taking man as model, in which case man is present in their designs precisely as a dead metaphor.

But if dead metaphors cannot, in themselves, lie at the origin of biomimetic innovation scenarios, what is their relevance to biomimetic innovation? Dead met-aphors are not in themselves capable of giving rise to biomimetic innovation, but there are two ways in which recognizing them may nevertheless provide a stimu-lus for biomimetic innovation. The first involves resuscitating the dead metaphor, bringing it back to life. Consider again the case of organic farming, which has changed quite significantly from Howard's initial conception of it and is today dominated by the idea of replacing synthetic chemicals with natural alternatives. Whereas Howard focused largely on imitating the trophic cycle of the forest, to the extent that biomimicry lives on in organic farming it is primarily via the idea of using natural materials—including both natural predators and pesticides of natural origin—for pest control, and it is thus primarily with respect to materials that it is biomimetic.[110] In view of this, it is clear that to resuscitate Howard's initial biomimetic metaphor of the "farm as forest" would open up the possibility for other imitable traits present in forests to be transferred to farms, including (to mention only a selection of traits discussed by Howard) nutrient cycling, soil aeration, and disease resistance. So, just as Kapp thinks that major technological breakthroughs emerge through unconscious return to their underlying model in the human body, technological or design breakthroughs in biomimicry may potentially occur through conscious resuscitation of dead metaphors.[111]

The other way in which recognizing dead metaphors may provide a stimulus for biomimetic innovation is through allowing us to see the dead metaphor in question as problematic, which may in turn prompt a search for a new "vital" metaphor. Identifying the dead metaphor of the "city as human organism," combined with understanding the various reasons it is problematic, could lead to the search for a more appropriate, vital metaphor, such as the "city as forest."[112] This in turn suggests that one of the reasons cities are not already like forests—and thus also why it is not a question here of biomimetic analogy or simile—is precisely that they have until quite recently been based on the quite different model of the human body or organism. Of course, there may also be other reasons why cities are not already like forests—perhaps because they resulted from the search for ideal solutions radically different from anything found in nature, or because they are not imitations at all, but rather free creations of the human mind not based on any preexisting models. But the very concept of dead metaphors also raises the intriguing possibility that, even when it was believed that we were discovering ideal solutions or freely creating, we were only ever inventing within a context created and dominated by the centuries-old dead metaphor of the *polisma anthropikon*, the city or state as human being. The anthropomimetic model of the human being may, in other words, have "ensisted" as a dead metaphor even when we believed we were discovering ideal solutions or freely creating cities independently of any model.

A final important topic to consider is the relative significance of the various innovation scenarios presented above. At present, only biological push and technological pull have been clearly identified in the literature.[113] The likely reason for this—that they often involve relatively simple and well-defined transfers between nature and technology—is also, I believe, their shortcoming. Indeed, in the case of both biological push and technological pull the danger not only arises that they will give rise to piecemeal solutions but also, and perhaps more significantly, that their "associated milieu," to use an expression of Gilbert Simondon, will be such that they will not call for, provoke, or intentionally seek to bring about the sort of generalized technological revolution that is required if we are adequately to respond to the ecological crisis.[114] Velcro, for example, has been seamlessly integrated into an associated milieu based to a large extent on the exploitation of petrochemicals in a linear cradle-to-grave economy. And while the Shinkansen improves energy efficiency, its effectiveness lies above all in its seamless adaptation to a preexisting technological milieu of train tracks and tunnels, rather than in its capacity disruptively to set the stage for the emergence of a new associated milieu.

In the case of biomimetic metaphor, by contrast, the latent stock of associated implications that may be transferred over from nature creates the possibility for multiple technological transformations to occur in parallel and in such a way that these transformations extend beyond the entities included in the metaphor so as to encompass their associated milieu. Indeed, this is precisely what occurs when Braungart and McDonough call on us not only to imagine buildings like trees but also to embed them in an associated milieu of cities like forests, thus extending the scope of the natural model present in the metaphor from the biological individual to the wider ecosystem. This is not, of course, to deny the importance of biological push and technological pull; particularly when they involve ecological models, they are very likely to make important contributions to ecological objectives. My contention is only that their revolutionary potential is not as great.

The revolutionary potential of biomimetic metaphor is also greater than that of biomimetic comparisons. In the case of biomimetic analogy and simile, there must already exist similarities between the natural entity and the technological artifact, in which case any innovations to which these scenarios give rise will generally be ameliorative, rather than revolutionary. But, since the principal problem we currently face is precisely that, generally speaking, human artifacts are radically dissimilar from natural entities, analogy will only be viable in those relatively limited or trivial instances where there already exist similarities between nature and technology. Likewise, there is, generally speaking, no obvious reason why the analogical amelioration of existing technologies is likely to be sufficiently disruptive to constitute a significant driver for radical ecologically based changes in their associated milieus. There may even be instances where biomimetic (and especially anthropomimetic) analogies or similes are highly inappropriate. Were one to notice, for example, that cities are in various respects analogous or similar to human or animal organisms, and, on the basis of that analogy, seek to make them even more like such organisms—for example, by installing information and communication networks modeled, however loosely, on nervous systems—one would ultimately be resuscitating a dead metaphor that, I suggest, is partly responsible for the current crisis, thus entrenching that crisis.[115] When, by contrast, we imagine radical transformations of our products and systems on the basis of vital biomimetic metaphors, we are not simply consolidating the dominance of existing products or artifacts through incremental ameliorations, but realizing what I contend is the biomimicry revolution's most profound and powerful innovation scenario.

## The Role of Chance in Biomimetic Innovation

A final important issue to address regarding biomimetic innovation concerns the role of chance. Just as Aristotle follows his analysis of the four causes by considering whether chance is also a cause, it is important for us to consider not just the types of natural causes, or better traits, that we may imitate and the different ways and scenarios in which we may imitate them, the role of chance in occasioning successful biomimetic innovation. In the present context, however, what is of primary interest is not the Aristotelian question of whether chance is a cause, for we may agree with him that it is, but rather the question of whether it is possible not so much to eliminate chance—for chance is, after all, a wonderful cause of innovation—as to increase the likelihood that we may chance upon biomimetic innovations.

In the classic cases of biological push and technological pull presented above—Velcro and the Shinkansen—chance clearly played a major role. Indeed, much of the charm of these two innovations is precisely their serendipitous nature. It so happened that a Swiss engineer came across the burdock thistle's hook and loop fastener while walking his dog and had the idea of basing an artificial fastener on it. And it so happened that a Japanese engineer designing a high-speed train also had an interest in ornithology and so could turn to the kingfisher's beak for a solution to his problem. Similarly, in the case of biomimetic metaphor, it would seem largely serendipitous that Braungart and McDonough formulated the seminal biomimetic metaphor of buildings as trees, cities as forests. The fact that Braungart is a natural scientist and McDonough an architect clearly played a role, but that simply raises the further question of why these two individuals seized upon the chance to collaborate with each other in the first place. Biomimetic analogy or simile is somewhat different. Since the two entities it brings together are already similar, all that is required is simply to notice that similarity and take it as a starting point for innovation. Relatively little imagination is required to imagine cameras that are more like human eyes (or like the eyes of other species, including those with compound eyes) than is for the most part currently the case.[116] But, as noted above, biomimetic analogy and simile are unlikely to play a leading role in the biomimicry revolution, for what this revolution requires is not so much an increase in existing similarities between natural and artificial entities as a revolutionary transformation of our artifacts so that their general dissimilarity from natural entities is overcome.

In view of the seemingly serendipitous nature of much biomimetic innovation, it has been suggested that transfers between nature and technology need to become "regularized" or "systematic."[117] In this context, much attention has been paid to the Russian problem-solving tool TRIZ. Developed by Genrich Altshuller from the study of thousands of patents, the aim of TRIZ is to facilitate the development of new innovations by helping identify viable solutions (from a stock of preexisting ones) as well as how to deal with any trade-offs involved. More recently, TRIZ has been extended to include biological solutions, thus giving rise to Bio-TRIZ, the basic aim of which is to systematize transfer from biology to engineering.[118] While Bio-TRIZ may perhaps have a significant contribution to make to the biomimicry revolution, there is one claim made by its advocates that I think is important to contest: that the choice we face is between the "systematic transfer" of models from biology to engineering made possible by Bio-TRIZ and the stark alternative of transfers that are "almost totally adventitious."[119]

The reason the choice is not this stark is that a middle ground already exists in the form of tradition. Just as works of art are not usually born out of either serendipitous flashes of inspiration or systematic transfers between empirical reality and human artwork, but occur instead in relation to a tradition, much the same is true in biomimicry. Indeed, within existing biomimetic research, there are already a number of important traditions within which innovation may occur. Some of these traditions concern level 1 models; certain pairings between natural models and technological counterparts are of such importance that they amount to traditions in their own right. Classic examples include self-cleaning surfaces based on the lotus plant, sticking pads making it possible to walk along vertical and overhanging surfaces based on the feet of geckos, and solar energy technologies based on natural photosynthesis. More commonly, however, traditions operate with respect to level 2 and 3 models, for it is here that one finds many of the major fields of biomimetic research within which the transfer of level 1 models may be subsumed. Examples of level 2 or 3 traditions include such fields as brain-inspired computing, nature-inspired chemical engineering, and agroecology.

Within all of these traditions, there is of course still scope for chance innovations, but it is also true that, in circumscribing pairings of relatively well defined natural and technological phenomena (or ensembles thereof), these traditions also circumscribe the sort of phenomena to be investigated. This is not to say that new traditions—or singular biomimetic designs—may not arise on the basis of chance leaps between areas of nature and technology previously considered unrelated or distant, or that existing traditions may not undergo revolutions on the basis of such leaps. My purpose in invoking tradition here is not to put an end to such leaps by constraining innovation to mainstream practices within existing traditions, but simply to show that such leaps are the exception rather than the rule, and that much innovation can take place in their absence.

It is also important to realize that much of what is required if the biomimicry revolution is to achieve anything like its full potential are not only new innovations but also, and perhaps more importantly, increased willingness and opportunities to work within existing biomimetic traditions. What is most important, for example, in the architectural tradition based on the biomimetic metaphor of "buildings as trees, cities as forests," is not so much that it yields radical technological breakthroughs as that it is widely adopted and perhaps even seen as a "new paradigm" in both architecture and urban planning and design. Common adoption of this metaphor would in turn create what one might, to use a biological analogy, call "selective pressure" in favor of relevant technological breakthroughs—for example, in artificial photosynthesis. But willingness and opportunity to adopt biomimetic metaphor in the first place, even if in so doing one does not generate any particularly original technologies or design features, is also required if the biomimicry revolution is to take place.

The question nevertheless remains: How might it be possible to increase the chance of successful biomimetic innovation? "Chance" may be defined as the beneficial or fortuitous encounter of two separate causal sequences.[120] In the case of biomimetic innovation, this corresponds to the idea of natural and technological knowledge and expertise fortuitously coming into contact in such a way that mimesis may productively occur. As a general rule, the way to do this is to reduce the distance between the two: the closer natural and technological knowledge and expertise are to one another, the greater the chance that a mimetically productive encounter may occur between them.

One obvious way to reduce this distance is, of course, through encouraging interdisciplinary collaboration, in particular between natural scientists, on the one hand, and, on the other, engineers, designers, architects, and the like, most obviously by changes to institutional structures. A related way of reducing the distance between natural and technical knowledge is what is referred to in biomimicry circles as "reconnection" with nature. Consider again the case of Velcro. Although I presented it earlier as a paradigmatic example of serendipitous discovery, it was only made possible because its inventor had gone for a walk in nature, thus, as it were, reducing the distance between his own knowledge and expertise in engineering and the natural knowledge present in the burdock thistle. Promoting reconnection with nature could take many forms, including increasing access to nature, bringing nature back to where people live (via urban restoration and rewilding projects, for example), or developing and extending education programs in natural science, especially for engineers, designers, and architects.

Other more theoretical and conceptual changes that may reduce the distance between natural and technological knowledge and expertise have already been presented above. The concept of technics, for example, makes it possible to bring together two areas of philosophical inquiry, philosophy of biology and philosophy of technology, that previously remained both marginal with respect to the basic structure of contemporary philosophy and distant from one another, thus creating a space for thinking about the mimetic relation between nature and technology. Similarly, the Aristotle-inspired typology of imitable traits facilitates the process of transfer between nature and technology, as does better understanding both the mimetic process (abstraction, transfer, and composition) and the innovation scenarios in which it may occur, especially the little-known scenarios of biomimetic comparison and metaphor. After all, if I have spent so long theorizing these theoretical and conceptual elements of biomimicry, it is, at a practical level, in large part because I hope that doing so will increase the chances of precisely the sort of productive encounters between natural and technical knowledge necessary for biomimetic innovation.

━━━◆━━━

An important but not always explicit thread running through the above exposition of biomimetic technics has been the aim of significantly expanding the scope of biomimicry. This has been achieved in five main ways. The first involves expanding our understanding of the different types of natural beings from which models may be abstracted, such that what is included are not only biological individuals but also certain physical beings, biological societies, and species, as well as Gaia, the one wholly natural ecological being. The second is the expansion of our understanding of the various traits we might imitate, such that these henceforth comprise not only natural forms (composite biotic materials included), but also the materials of which natural systems are composed, the processes that generate them or that give rise to certain effects, and the functions they carry out. The third involves the idea of imitating nature at higher levels of abstraction—that is to say, not only at the level of traits specific to some natural being or other, but also at the level of more general and abstract traits often common to multiple natural beings. The fourth involves recognizing that various differences between natural systems and biomimetic technologies may be introduced within the process of biomimetic innovation, through the processes of translation, transformation, and composition. And the fifth involves not expanding our understanding of the logical space of possibilities within which nature imitation may occur, but, rather, expanding our awareness of the different innovation scenarios that may arise within that space, including the hitherto unrecognized scenarios of biomimetic comparison and biomimetic metaphor.

A very significant consequence of this expanded scope is the *normalization* of biomimicry. As it stands, most people currently view conventional technology as normal and biomimicry as the exception. But, in view of the expanded scope of biomimicry presented above, it would seem that it is biomimicry that is normal and conventional technology that is the exception. Indeed, the overall space of technological design is that of biomimicry, with conventional technology being rather what philosophers of science and mathematicians call a "limiting case," a case that may be included within the overall theoretical framework but has the particularity of pushing the values possible within that framework to their logical extreme, such that, in the present instance, *no trait of any natural being is imitated, no matter what the level of abstraction, and no matter how creatively.* In every other case, imitating nature is involved, and the technology in question falls within the standard set of cases theorized under the rubric of biomimetic technics. In this manner, biomimicry becomes normalized, with conventional technology henceforth being seen as an exception characterized by the extreme values assumed by all the relevant parameters.

This normalization of biomimetic technics is, I believe, highly significant for our understanding of both the history and the philosophy of technology. Writing in *Nature*, the popular science writer Philip Ball speaks of what he calls the "long and colourful history" of engineers and the like taking inspiration from nature—a history he illustrates with examples including Joseph Paxton's design for London's Crystal Palace of 1851 (inspired by the ribbed stem of the lily leaf) and René Binnet's entrance gate to the World Exposition of Paris in 1900 (inspired by Ernst Haeckel's drawings of radiolarians).[121] While not exactly false, Ball's way of characterizing the history of nature imitation is hugely misleading, for it overlooks the fact that human thinking about nature imitation has undergone various periodic changes and revolutions, some of which may even be obscured by the isolated examples he presents. In particular, Ball's approach misses three key points: first, in the modern period—especially from the nineteenth century onward—the long-standing idea of technē as imitation of nature, which goes back to the ancient Greeks, was widely rejected; second, this rejection coincided with the emergence of explicit philosophical thinking about technology in the work of Kapp, Eyth, Dessauer, and others; and, third, all of this coincided with the industrial revolution, and thus also with the proximal beginnings of the environmental crisis unfolding ever more alarmingly today.

If, however, it is ultimately nature imitation that is "normal," then that relatively short phase of Western history, during which technology strived to attain "spiritual autonomy" from nature, would come to appear as an historical exception, and philosophy of technology, which, at least as a recognized area of

philosophical inquiry, began with and has only ever known the "spiritual autonomy" of technology, would face an existential challenge. What allowed "philosophy of technology" to exist in the first place was the way it isolated technology
as an object of study, cutting it off from the natural models on which it had
previously depended. Human technics was thus cut off from natural technics,
with the consequence that human designs were no longer seen as deriving from
natural models, but from the minds of humans, with nature at most providing
only the energy and raw materials necessary for the concrete realization of these
designs. But if, as I have proposed, these two branches of technics—the natural
and the human—are to be reunited, with the former once again assuming the
role of model for the latter, then the question arises as to whether "philosophy of
technology" can even continue.

The normalization of biomimetic technics is also of great practical significance
for the work of designers, engineers, and the like. Beyond the provision of a comprehensive framework for understanding the process of biomimetic innovation,
the theory of biomimetic technics set out above also opens up the possibility for
fields of innovation currently seen as quite separate or distinct to fall under its
scope. The field of biodesign, for example, which is often theorized as distinct
from biomimicry on the grounds that it involves the *integration* of living beings
into human designs, as opposed to taking them as *models* for human designs, could
nevertheless embrace biomimicry by coming to see the integration of living beings
as itself biomimetic, inasmuch as it follows nature in using living beings as biotic
materials.[122] Similar points apply to such fields as the bio-based economy and synthetic biology. In this manner, the framework for biomimetic innovation may
acquire practical significance beyond the relatively narrow field of biomimicry
(e.g., biomimetics, bioinspiration), as it is currently understood and practiced, and
would ultimately come to be understood as a general framework covering all technological innovation, conventional innovation included, albeit only as a limiting
or special case.

This would in turn have significant pedagogical implications. As it stands, biomimicry is seen as a marginal approach to innovation, albeit one whose importance is increasingly recognized and appreciated, especially in the context of the
current quest for sustainability. In keeping with this, biomimicry typically occupies a marginal role in the education and training of engineers, designers, architects, and the like, for it is at best taught as an optional course within a broader
program of study, the core of which still presupposes technology's "spiritual

autonomy." But if imitating nature were once again to be seen as the "norm" for technological innovation, this would need to change; biomimetic technics would henceforth contribute significantly to the formulation of core programs of study in the technical disciplines (e.g., engineering, design, architecture). In practice, this would likely involve increased training in the life sciences, increased training in how knowledge in the life sciences can be abstracted and transferred over to the technical disciplines, and, as a consequence of these two changes, greater willingness on the part of those working in technical disciplines to collaborate with colleagues in the life sciences.

## CHAPTER 3

## *Nature as Measure*

Biomimetic Ethics

T here can be little doubt that biomimicry requires an ethics. The discussion of the principle of nature as model in the previous chapter gave us an indication of the endless ways in which we may imitate nature, but it set out no ethical norms to guide that process. This raises the question of whether we may simply identify some preexisting ethics and apply it to biomimetic innovation, or whether, to the contrary, biomimicry already contains an ethics, or at least the latent possibility of an ethics, in which case an ethics may arise internally to biomimicry and in a way that is singularly appropriate to it. Embracing this latter perspective, the present chapter locates this ethics, or at least the possibility of its explicit development, in the second basic principle of biomimicry as set out by Benyus: nature as measure. Nature, from this perspective, provides not only models we may imitate but also a measure against which we may judge the ethicality of our innovations.

That an ethics might be developed from the principle of nature as measure also raises the further possibility that this ethics may have a wider scope than simply setting constraints on innovation. The principle of nature as measure may, in other words, underpin more than just an ethics of technology, for it may also provide the basis of an ethics applicable to other spheres of human action. Nature as measure would thus come to stand alongside nature as model as a basic principle in its own right, and not simply as an ethical adjunct to the principle of nature as model. In so doing, it would ground not only what one might call an "ethics of biomimicry," in the sense of an ethics applicable to biomimetic innovation, but also a *biomimetic ethics*—that is to say, a new ethics developed from out of the basic philosophical framework of biomimicry and that has a wider sphere of application than just technological innovation.

The principal aim of this chapter is to present and develop just such a biomimetic ethics. Before doing so, however, let us first consider its revolutionary character. To take nature as measure, as Benyus proposes, is to introduce a radically new principle into ethics. At the same time, this principle also suggests the possibility of developing a historically structured typology of different ethical systems, according to which what fundamentally distinguishes them from one another is the entity or entities in which they locate the measure of right action. In parallel, then, to the history of *models* other than nature discussed at length at the outset of the previous chapter, we may consider the history of those *measures* other than nature that have, over the course of Western history, formed the successive bases of our ethics: God in the case of theomimicry; Man in the case of anthropomimicry; and anything and everything in the case of pantomimicry.

Just as theomimicry has a technical principle, God as model, it also has an ethical principle, God as measure. God's way of being sets the standard against which human beings and their actions are to be judged. God is infinitely benevolent, loving, just, and so on, and it is this divine standard that human beings are to take as measure. Certain exceptional human beings, such as saints, may approximate to the divine standard, and so become concrete exemplars for lesser mortals to follow, but even they could never fully live up to the moral perfection of God Himself. Further, God's moral perfection extends not only to His way of being but also to the act of creation, which was not a mere exercise in virtuoso craftsmanship, an act of incomparable and unsurpassable skill, for it proceeded ultimately from His benevolence, from the incomparable and unsurpassable goodness of his will. There was, in short, an ethical or moral basis to creation.

Like theomimicry, anthropomimicry has an ethical principle, man as measure, in addition to the technical principle of man as model. From this perspective, man's essential way of being provides the standard against which human actions are to be judged. An obvious example is the ethics of Kant. For Kant, the human essence lies in reason, and a truly rational action will ipso facto be an ethical one; reason requires us to universalize the maxim underlying our action, and, in doing so, we necessarily treat others as ends and never solely as means, thereby avoiding such unethical actions as killing, stealing, lying, or breaking promises.[1] The standard set by human nature is not, however, one that could ever be fully met. Ethical action will always be tinged, Kant tells us, by nonrational passions and affects; whenever we act in accordance with morality we will always be driven to do so at least in part by feelings, such as compassion or sympathy, and our action will thus lack the moral purity of the rational standard.

In more recent times, the principle of man as measure characteristic of Enlightenment thinking has been called into question. In the wake of Friedrich Nietzsche's call for the reevaluation of all values,[2] existentialists affirmed with Jean-Paul Sartre that "existence precedes essence," in which case there is no preexisting human essence that can provide the standard against which human existence can be measured; our essence is freely created.[3] So, just as on Blumenberg's reading, twentieth-century philosophy of technology began to see man as the creator of entirely new technologies, so existentialism affirmed man's ability to create entirely new values, entirely new standards for evaluating right action. Parallel to this, there also occurred the transition from objective to subjective reason theorized and critiqued by Max Horkheimer.[4] Whereas objective reason holds that standards for right action can be derived from human nature, in which case there is some objectively valid reason to act in some ways and not others, subjective reason sees the sole reason for action as emerging from the subjective desires or preferences of the individual.

Lastly, and in keeping with the rise of the "pantomimetic" principle of everything (and anything) as model characteristic of postmodernism, there also arises the principle—expressed notably by Bruno Latour—of everything (and anything) as measure of everything (and anything) else.[5] Our primary ethical duty thus becomes to stop any one thing rising up and establishing itself as the supreme measure for right action, to stop any attempt to reduce the irreducible plurality of standards to one.[6] From these post-Christian and posthumanist perspectives, there is no supreme measure—deriving from the nature of God or Man—that all must respect, but rather a plurality of standards emanating from multiple different beings.

With the biomimicry revolution, a new measure of right action presents itself: nature. The rightness of our actions—or at least some of them—is to be judged against nature's ecological standard. Like any basic principle, however, this principle clearly requires interpretation, and, in the act of interpretation, a number of fundamental philosophical questions emerge. The first of these concerns whether the principle is an ethical one, for, while clearly normative, it is not immediately apparent why it is *ethically* normative as opposed to just *technically* normative. Or, to put it another way, why would one suppose that it contains primary ethical norms guiding how we are to live, rather than just secondary norms guiding the technical performance of the artifacts we build? This in turn raises a further question. Even if the principle is ethically normative, to what sphere does it apply—technology, interhuman relations, or perhaps rather the human relationship to nature? And if, as I will argue, it applies only to the latter—that is to say, to environmental ethics—how exactly does it relate to this field as it currently exists? Does the principle of nature as measure lead to a new *position* in environmental ethics, or does it go further and introduce an entirely new *approach* to environmental ethics? And, if it is the latter, what would that approach, and any specific theoretical position one might adopt on its basis, contribute to current debates in environmental ethics, including, in particular, the famous conflict between anthropocentrists and nonanthropocentrists about whether our ethical duties are solely to humans, or also to nature?

## ECOLOGICAL AND ETHICAL STANDARDS

### Nature as Measure

With a view to answering these questions, let us being by considering Benyus's own explication of the principle of nature as measure. This explication comprises two parts. The first part is simply affirmative: "Biomimicry uses an ecological standard to judge the 'rightness' of our innovation."[7] The reference to "rightness" immediately allows us to see that, whereas the principle of nature as model is strategic, the principle of nature as measure is normative; it places normative limits on the otherwise limitless possibilities opened up by the design strategy of taking nature as model. It is also significant that the standard against which rightness is measured is not characterized simply as "natural," as one might expect given the principle of "nature" as measure, but rather as "ecological"; it is one region of nature that provides the standard, ecology, and not, as it were, nature in general (i.e., physis on the present theorization of nature).

The second part of Benyus's explication goes on to provide a justification of this affirmation: "After 3.8 billion years of evolution, nature has learned: What works. What is appropriate. What lasts."[8] This justification may also be divided in two. First, it tells us that nature has acquired knowledge of three things that, so it would seem, are self-evidently valuable. The first of these, what works, may be identified with effectiveness (or efficiency); what works is what is effective (or efficient). The second, what is appropriate, may be identified with fittingness; what is appropriate is what is fitting. The key feature of fittingness is that it is contextual; what is appropriate or fitting in one context may not be so in another. The third, what lasts, may be identified with sustainability. What lasts is what may endure or be sustained over long periods of time, perhaps indefinitely. So, whereas fittingness introduces a contextual dimension to the knowledge embedded in nature, such that it is always adapted to a specific place or context, sustainability introduces a temporal dimension, specifically the capacity to endure.

The self-evident character of these three forms of knowledge embedded in nature becomes manifest when one considers that their opposites—the ineffective, the inappropriate, the unsustainable—should in every case be avoided. Evaluation against nature's ecological standard would thus appear to be justified on the grounds that it involves judging our innovation against three self-evidently valuable forms of knowledge embedded in nature: knowledge of the effective, the fitting, and the sustainable.[9]

Benyus also informs us of the reason that nature has been able to acquire these self-evidently valuable forms of knowledge: it has had an unimaginably long time to do so, approximately 3.8 billion years. This is not, however, intended simply as an argument from venerability; the knowledge acquired by nature does not function as a measure because nature is very old. The argument is, rather, that evolutionary processes tend over long periods of time to select the effective over the ineffective, the appropriate over the inappropriate, and the sustainable over the unsustainable. As a result of its well-known "blindness," evolution may be slow in comparison with the cultural learning characteristic of humans, but over very long periods of time it will nevertheless tend to lead to the accumulation of these three self-evidently valuable forms of knowledge. So, while we humans may have learnt a great deal, particularly over the last decades and centuries of rapid technological, economic, and social development, including, no doubt, many things unknown to nature, the proposal to take nature as measure implies that, as a general rule, we currently differ from nature in that the knowledge we have been generating is not simultaneously effective, appropriate, and sustainable. Contemporary human knowledge may in some instances be highly effective, at least as measured by conventional standards of technological performance (e.g., speed, durability, or accuracy), but not only is it in some cases less effective than nature's, more importantly, it is also typically much less appropriate and sustainable.

In light of the above analysis of the structure, content, and rationale behind the principle of nature as measure, let us now consider its philosophical status, and more specifically the question of whether the *ecological* standard it upholds is also an *ethical* standard. Now, effectiveness (or efficiency) has long been seen as the fundamental norm of technology. If, as is generally assumed, technology is a means to an end, then it must achieve its ends in ways that are both effective (i.e., that work) and efficient (i.e., that are not wasteful). This is why, among those recent and contemporary philosophers of technology who concentrate on the analytic issue of precisely defining technology, effectiveness and efficiency typically take center stage, the result being that theoretical differences between these philosophers often come to depend above all on subtle variations regarding the basic norm or ideal of effectiveness or efficiency. James Feibleman, for example, claims that "the ideal of technology is effectiveness."[10] I. C. Jarvie emphasizes the epistemological dimension of technology, arguing that technological knowledge is "knowledge of effectiveness."[11] Mario Bunge adds that the distinctive feature of technology, as opposed to arts and crafts, is that through reference to science it can account for *why* it is effective.[12] And Henryk Skolimowski agrees that technological knowledge is of effectiveness, but adds that within the category of effectiveness one must include a wide range of other criteria, like durability, resistance, sensitivity, and speed, depending on the entity in question.[13]

But, if the fundamental norm or ideal of technology is effectiveness (or efficiency), how does technology relate to the norms of fittingness and sustainability? Skolimowski's understanding of effectiveness, according to which this category may be further subdivided into other categories, suggests the possibility that fittingness and sustainability could potentially be integrated within traditional philosophy of technology as additional subcategories of effectiveness. The problem with this approach is that, unlike such subcategories as speed, durability, or accuracy, which are only normative in certain cases, fittingness and sustainability are normative in *all* cases. So, whereas speed may be normative for planes but not for chairs, durability for bridges but not for biodegradable packaging, and accuracy for thermometers but not for wheels, the self-evident value of appropriateness and sustainability means that their normativity, like that of effectiveness, assumes a fundamental and universal character. Just as all technologies should be effective, so all technologies should be appropriate and sustainable. Fittingness and sustainability thus differ from such properties as speed, durability, and accuracy in that they cannot be reduced to subcategories of the effective, but must instead, as is the case in Benyus's explication of the principle of nature as measure, sit alongside effectiveness as self-evidently normative criteria for judging the rightness of our innovations.

But if, as philosophy of technology has traditionally maintained, the sole grounding norm for technology is effectiveness (or efficiency), and if appropriateness and sustainability cannot be reduced to subcategories of the effective (or efficient), the possibility presents itself that the norms of appropriateness and sustainability are not technical norms at all, but, on the contrary, ethical norms that place limits or constraints on the technical norm of effectiveness (or efficiency).[14] Pursuing effectiveness (or efficiency) at the expense of appropriateness and sustainability would thus count as unethical. With a view to exploring this possibility, let us now turn our attention to the semantic proximity between the Greek words that lie at the origin of our words "ethics" and "ecology"—namely, *ethos* and *oikos*.

## Ethos and Oikos

The word "ethos" has a variety of meanings. The first is "character," which is today the standard translation of *ethos*. The second is "custom," which is roughly equivalent to the Latin *mores*, a word that has given rise to our contemporary word "mores." And the third is "place." In the "Letter on Humanism," Heidegger insists that it is this latter meaning that is the authentically Greek one, and not "character," as is usually presumed.[15] This in turn leads him to suggest that the study of ethics is ultimately the study of the human home or abode, the place where we dwell.[16] But, while Heidegger is right to underline the fundamental semantic link between *ethos* and place, this should not lead us to believe that to translate *ethos* as character is wholly mistaken, as if ethos had nothing to do with character. If *ethos* can have the meanings of character, mores, and place, it is precisely because these three meanings are related. The basic meaning of *ethos* differs from the usual Greek word for place, *topos*, in that it refers not to place in general, but, rather, as Heidegger has shown, to an inhabited place, and more specifically to the home or abode of humans. But humans cannot dwell without common mores that structure and regulate their dwelling. And these mores in turn condition the standards by which goodness of character or virtue is measured. From this perspective, ethics is more than just the study of place, and more even than the study of the abode of man; it is the study of the relation between the place where humans dwell, their mores, and their standards of good character.

Grounding ethics in place is anathema to modern philosophy. Modern philosophy grounds ethics in some intrinsic feature or other of human beings, thus overlooking any contextual relationship to place. For Kant, ethics is rooted in rationality, and since rationality bears no relation to place, it follows that ethics likewise bears no relation to place. All rational beings, including, were they to exist, aliens inhabiting completely different regions of the universe (or even different universes), would have to obey the same moral code, not because of any common relation to place, but simply because of their intrinsic nature as rational beings. Likewise, for utilitarianism, ethics is rooted in the human (and animal) capacity to experience pain and pleasure, but this capacity bears no intrinsic relation to place. Even virtue ethics—at least prior to the emergence of environmental virtue ethics—has lost sight of the original connection between virtue and place, becoming instead a theory of the intrinsically valuable moral traits of human beings, regardless of where it is that they dwell.

It could of course be objected here that there is good reason not to connect ethics to place, for it robs ethics of its universality, implying a problematic cultural relativism. It was from precisely this perspective that Emmanuel Levinas put forward some of his most violent criticisms of Heidegger.[17] Heidegger's ethics, Levinas claimed, was rooted in place, whereas his own was a universal one—detached from the relativism implicit in place-based ethics—and thus opposed to the "scission of humanity" into "natives" (*autochtones*) and "foreigners" (*étrangers*).[18] This is why Levinas extolls Yuri Gagarin, the first human being to have left all place behind and to have existed, even if only temporarily, in the "absolute of homogenous space."[19] In keeping with this, modern philosophers would no doubt also be wary of the way the ancient concept of ethos appears to combine two notions that later came to be sharply distinguished: morality or ethics, on the one hand, and mores or customs, on the other. While the study of morality and ethics is concerned with providing right action or good character with rational justification, thus making possible universal assent, mores and customs emerge more or less unreflectively within the context of specific places, and, as such, could at best stumble upon the true ethical norms that philosophical reflection on the universal alone can uncover.

With a view to responding to this objection, let us begin by noting that there may be places within places. It follows that there may exist one single global place within which multiple local places may also exist. But if there is one single global place that encompasses all the others, ethics may be fundamentally the same for all humans and yet at the same time relative to place, for it may be relative to the common abode or dwelling place of all human beings. It is not hard to see, moreover, that this was precisely Heidegger's position. For Heidegger, ethics is concerned primarily with what he calls the "abode of man," the place inhabited not by individuals, cultures, or peoples, but by humanity in general. Within this primary global place, there will of course be secondary local places, and thus also local differences both in mores or customs and in standards of good character. But these local differences would be rooted in a shared ethos common to all human beings qua inhabitants of the one and only global place, in which case ethics would be concerned primarily with understanding this singular global ethos and only secondarily with its local variations or derivatives. Ethics, from this perspective, would still be relative to place, but it would not be relative to local culture, for it would hold for all human beings inasmuch as they all inhabit the same global place. Indeed, from this perspective, it is precisely the rooting of ethics in a shared global place—and not in some allegedly universal feature of the individual human mind—that would make it possible to overcome the cultural relativism of ethics, and therewith also the schism of humanity into "natives" and "foreigners," for we are all natives to one and the same global place.

It is the place-based nature of this view of ethics that connects it to ecology. Ecology, as its etymology tells us, is the study of the oikos, the home.[20] Both ethics and ecology, it follows, are concerned with inhabited places. There is, however, an important difference between the ethos of ethics and the oikos of ecology. In the case of the former, the inhabitants in question are human, whereas in the case of the latter, the inhabitants are living beings in general. This difference does not, however, present a fundamental barrier between the two; instead, it opens up the possibility that the oikos of ecology, the home of living beings, may provide the standard against which the home or dwelling place of humans, the ethos of ethics, may be judged. Just as nature and technology differ in that the latter is specific to humans, and yet, when both are understood as complexes of the same four causes (or basic traits), it becomes possible to take the former as providing models for the latter, so oikos and ethos differ in that the latter is specific to humans, and yet, when both are understood as "inhabited places," it becomes possible to take the former as providing measures for the latter. It is, in short, the discovery of a fundamental conceptual parallel between oikos and ethos, between ecology and ethics, that makes it possible for the former to provide the latter with its fundamental measure.

If, as the principle of nature as measure requires, we take the oikos as providing the measure of the ethos, this implies that the values and norms of appropriateness and sustainability are indeed ethical in nature. The concept of appropriateness, as we saw earlier, is inherently contextual, inherently related to fitting in to place. But so also, as we have just seen, is the concept of ethics. Appropriate inhabitation of place is at the same time also appropriate ethics. Further, in keeping with the idea that there is one single global place containing various other local places, to root ethics in an ecological standard is not to root it in one local place at the expense of other local places, but rather in a single global place common to us all, humans and nonhumans alike: the earth.

As for sustainability, it relates rather to the concept of customs or mores, the second key component of the Greek concept of ethos. Customs and mores are not one-off actions, but ways of being that have stood the test of time. The problem we currently face, however, is that our current ways of being, though they may have increased prosperity in the short-term, are unsustainable and so, by definition, cannot stand the test of time. To see nature as providing a standard not only for how we may adapt and fit into the places we inhabit (both global and local) but also for the development of sustainable ways of being, and therewith also new customs and mores, is thus to see nature as providing a new standard for ethics. It is also important to realize that, while nature exhibits great variety and undergoes constant changes, it may nevertheless be possible to abstract from her overall way of being a certain number of general traits that have stood the test of time and that, as such, may provide a template for a new set of customs and mores.

With this in mind, it is instructive to note that Benyus puts forward a "canon"—a word whose etymology refers to a standard or measuring rod (from the Greek, *kanon*)—of the "laws, strategies, and principles" she considers characteristic of nature:[21]

> Nature runs on sunlight.
> Nature uses only the energy it needs.
> Nature fits form to function.
> Nature recycles everything.
> Nature rewards cooperation.

Nature banks on diversity.
Nature demands local expertise.
Nature curbs excesses from within.
Nature taps the power of limits.[22]

Quite which of these are supposed to be laws, which strategies, and which principles is unclear. It is also unclear on what basis exactly this list has been established. It may even turn out that some of the items on this list are problematic, that Benyus is mistaken, or at least partially mistaken, about nature's ways of being. But the general point remains that it may be possible to abstract certain common and enduring traits from nature and to use these as the basis for the establishment of new and sustainable customs and mores.

### Biomimetic Ethics as Environmental Ethics

An important question raised by the idea, constitutive of biomimetic ethics, of deriving an ethics from the principle of nature a measure concerns its sphere of application. According to Jackson, who, as we saw in the introduction, first proposed the principle of nature as measure,[23] this measure does not apply to interhuman ethics.[24] In particular, the ways that nonhuman living beings relate to each other within ecosystems—and that often involve such phenomena as predation and parasitism—are not to be taken as measures for interhuman relations.[25] The obvious objection that taking nature as measure implies socio-Darwinism, or something similar, is thus circumvented.[26]

This is not to say that the idea of rooting ethics in place could not also lead us to rethink interhuman ethics. More specifically, interhuman ethics could be seen as rooted in fundamental structural features of interhuman relations as they arise and occur within a shared dwelling place common to humans in general, and not, as is typically the case today, in some placeless property of the individual human mind, such as reason or sentience. But such a standard would not be biomimetic, in the sense that it would not derive from nature, but from fundamental structural features of interhuman relations as they arise in a shared place *specific to humans*, and that, in the previous chapter, I followed Heidegger in calling and understanding as the "clearing." An ethics of the clearing, it follows, would not be solely biomimetic; the interhuman aspect of that ethics would not derive from nature, but from those structural features of the clearing that relate to human relations (i.e., from what Heidegger calls the "openness of man to fellow man"). Since that interhuman aspect would not be biomimetic, however, it falls outside the scope of the present work.

If biomimetic ethics does not concern interhuman ethics, it could perhaps be thought that it concerns only our relation to technology, that it is ultimately just an ethics of technology. With a view to evaluating this claim, let us begin by noting that, at least since Aristotle, the spheres of ethics and technology have generally been kept quite separate. For Aristotle, there is a sharp distinction between *praxis* (action) and *poiēsis* (production).[27] The telos of praxis, he thinks, is to realize the being of the agent; one acts in order to be fully human, to realize human excellence. The telos of poiēsis, by contrast, is to realize a being other than the agent, to bring *its* being to a state of completion. To these different objectives or *teloi* there correspond different types of knowledge: *phronesis*, or practical wisdom, is the type of knowledge that guides praxis; and technē, or know-how, is the type of knowledge that guides poiēsis. This distinction between praxis and poiēsis in turn allows ethics and technics to be treated quite separately: ethics is the study of the standards of good character that provide the normative basis for action and that phronesis helps us respect; technics is the study of the artful combination of various different causes with a view to producing an artifact.

The problem with this radical distinction lies not in the fact that it distinguishes between poiēsis and praxis, between technics and ethics, but rather in the fact that it has led the study of technics and the study of ethics to be pursued quite independently, such that the relation between the two—and in particular the question of how it is that ethical norms might guide the practice of producing artifacts—has not traditionally been addressed. For Aristotle, the production of artifacts, such as beds, tables, bronze statues, and silver chalices, was not a matter of ethical concern, but simply of skill. There was, for him, no ethics of technology. But this is to ignore the fact that technē (poiēsis) is also a form of action (praxis), inasmuch as the production of entities other than the self may or may not allow us to realize fundamental ethical ends—human excellence (*aretē*), in Aristotle's case. In keeping with this, we today are increasingly aware that while skill or know-how is certainly required to produce artifacts, many questions concerning what we produce and how (e.g., whether to produce energy from solar panels or from coal-fired power stations) are of great ethical concern, for how we answer them will play a great part in determining to what extent the earth will remain habitable. Technology, it follows, cannot simply be left to its own devices, but must instead be limited and constrained by a viable theory of ethics.

To see biomimetic ethics as simply an ethics of technology would, however, be to give it a scope that is at once too broad and too narrow. It is too broad inasmuch as technology does not only mediate the human relationship to nature but also interhuman relationships, and, since biomimetic ethics is not concerned with interhuman relationships (or at least not directly), it is not relevant in that context. If the experimental testing of some new technology on humans were to be prohibited on ethical grounds, it would not be because of nature's ecological standard, but because some essential feature of "man's openness to fellow man" opposes it. Further, the simple fact that a technology is biomimetic does not mean that its rightness or wrongness may be determined by biomimetic ethics. When humans are replaced in their economic functions by biomimetic technologies developed in such fields as AI and robotics, it is not nature's ecological standard that allows us to determine whether this is right or wrong, but interhuman ethics.[28]

And yet to see biomimetic ethics as simply an ethics of technology is also too narrow, for there are various spheres of action other than technological innovation where nature's ecological standard may also be applied. There is, from this perspective, simply no reason to restrict the respect of ecological norms derived from nature to technological innovation. Indeed, such norms may apply not just to the production of technologies, but to all sorts of other actions as well. Two general types of action are of particular importance. The first are everyday economic activities, such as shopping, eating, and traveling, which, though permeated and enabled by technology, are nevertheless quite different from technological innovation and production (poiēsis), for they do not involve the gathering of the four causes to produce an artifact. The second are those actions that directly concern the environment, especially preservation, conservation, and restoration, to which, as we will see later, the principle of nature as measure may also be applied. There can be little doubt, then, that to live according to ecological standards does not only concern production (poiēsis), but also action (praxis). Being ecological, we may conclude, is not simply a matter of technological innovation, but of other types of action as well.

But if biomimetic ethics does not directly concern interhuman ethics, and if it is not simply an ethics of technology, to what sphere does it apply? The answer,

I suggest, is that it is an ethics that governs the human relationship to nature, that it is, in short, an environmental ethic. Whether it is a question of technological innovation, everyday activities, or environmentally oriented action, the measure provided by nature concerns the question of how it is that we are to inhabit the earth, how it is that we are to live with respect to nature. With this in mind, let us now consider the relation between biomimetic ethics and environmental ethics.

## ENVIRONMENTAL ACTION ETHICS

### *Three Approaches to Environmental Ethics*

In order to prepare the ground for a discussion of the relation between biomimetic ethics and environmental ethics, I will first describe briefly the main approaches and positions within environmental ethics, before going on to argue that there is another approach—vital for the philosophical contextualization of biomimetic ethics—that has yet to be explicitly theorized: environmental action ethics.

Environmental ethics emerged as a recognized academic discipline in the 1970s and 1980s through the work of such pioneering philosophers as J. Baird Callicott, Val Plumwood, Holmes Rolston, Richard Sylvan, and Paul Taylor. Its central question has thus far concerned who or what is to count as an *object* or *patient* of ethics—in other words, which beings have so-called moral standing, with the answer depending on the identification of a morally relevant property, specific either to the beings in question or to the relationship we have with them, and from which certain duties are held to follow.[29]

Broadly speaking, four different positions have been put forward within environmental ethics (see table 3.1). Anthropocentrism, which is also the traditional position in Western ethics from Plato and Aristotle onward, affirms that only humans have moral standing, usually on account of their putatively unique ability to reason or to speak. In the context of environmental ethics, anthropocentrism calls for a wider appreciation of nature's utility, thanks notably to a better understanding and appreciation of the ecosystem services it provides; a consideration of a wider set of human beings, such that future generations are also taken into account; and, in the case of weak anthropocentrism, significant changes to our values and preferences, which would no longer center around consumption.[30]

But, even after these changes, human beings remain the ultimate beneficiaries of environmental action and policy, even if many nonhumans would benefit indirectly.

Opposed to anthropocentrism, there exist three main forms of nonanthropocentrism. Zoocentrism affirms that all animals, or rather all beings endowed with sentience,[31] have moral standing, for it is sentience—either in itself or in the form of its positive contents (pleasure, absence of suffering)—that has intrinsic value.[32] Biocentrism affirms that all living beings have moral standing, usually by virtue of the argument that they are ends in themselves, because, unlike mere tools, which are only good to the extent that they serve some other end, living beings have a "good of their own" manifest in the various ends they pursue (e.g., sustenance, reproduction).[33] And, lastly, ecocentrism affirms that it is possible to expand the moral community to which we feel or believe we belong such that it comes to include not only other living beings but also various abiotic elements, such as soils and waters, with our moral duty being ultimately to preserve various systemic features of that community—namely, at least in Aldo Leopold's formulation, its integrity, stability, and beauty.[34]

More recently, another approach to environmental ethics has emerged: environmental virtue ethics. In contrast to mainstream environmental ethics, environmental virtue ethics focuses not on the being of the moral object (or moral patient) and the duties that derive from it, but rather on the being of the moral subject (or moral agent) and the virtues it may come to possess. The key question, in other

TABLE 3.1   Positions in Mainstream Environmental Ethics

|  | Anthropocentrism | Zoocentrism | Biocentrism | Ecocentrism |
|---|---|---|---|---|
| **Object of moral duty** | Human beings | Animals | Living beings | Ecosystems, the biotic community |
| **Morally relevant property** | Logos, reason | Sentience | Self-directed telos | Moral sentiments toward the biotic community |
| **Nature of duties** | Reciprocal moral obligations | Maximize utility / respect subjectivity | Protect and promote the good of living beings | Preserve systemic properties of the biotic community |

words, is not what beings have moral standing, but what are to count as virtuous traits, attitudes, and dispositions with respect to nature on the part of the subject. In some cases, environmental virtue ethics may correspond quite closely to mainstream object-centered environmental ethics. In Taylor's classic statement of biocentrism, for example, there exist what he calls "special virtues" corresponding to the various duties he thinks we have toward living beings; the virtue of considerateness corresponds to the duty of nonmaleficence (i.e., not harming other species), the virtue of impartiality (between species) to the duty of noninterference (in the lives of other species), and so on.[35]

Other formulations of environmental virtue ethics provide a broader understanding of environmental virtue. Ronald Sandler, for example, sets out a typology of environmental virtues, which are divided into five overarching categories (see table 3.2). Of these, only those of respect for nature correspond to the sorts of duties toward nature with which mainstream object-centered environmental ethics is concerned. So, while an attitude of considerateness toward nature may correspond to a moral duty (nonmaleficence, on Taylor's view), the virtues of communion with nature, environmental activism, or environmental stewardship do not correspond to moral duties toward nature. It may be a virtue to have a heightened sense of wonder in the face of nature, or to engage in environmental protests and campaigns, but it is not obviously a moral duty.

Therefore, just as traditional virtue ethics seeks to describe and promote directly other-regarding moral virtues, such as compassion or justice, as well as general virtues that tend to benefit both self and others, such as prudence, temperance, or courage, the same is true of environmental virtue ethics.[36] There is thus a partial but not total overlap between mainstream environmental ethics and environmental virtue ethics, with the former concentrating on the characteristics of natural objects from which moral duties derive, and the latter on a broader set of environmental virtues than those which correspond to moral duties toward nature, including general virtues of benefit to both nature and human agents.

The very fact that there exists an alternative to mainstream environmental ethics— namely, environmental virtue ethics, the key idea of which is that environmental ethics is centered not on the moral *object*, but on the moral *subject*, raises the theoretical possibility that there may exist yet another approach to environmental ethics—one whose center lies elsewhere. And this third center, I suggest, is none other than the *environmental action* that mediates the relation between

TABLE 3.2　A Typology of Environmental Virtues

| Virtues of sustainability | Virtues of communion with nature | Virtues of respect for nature | Virtues of environmental activism | Virtues of environmental stewardship |
| --- | --- | --- | --- | --- |
| Temperance | Wonder | Reverence | Diligence | Benevolence |
| Frugality | Openness | Compassion | Cooperativeness | Loyalty |
| Far-sightedness | Appreciation | Restitutive Justice | Commitment | Justice |
| Attunement | Attentiveness | Considerateness | Optimism | Honesty |
| Humility | Love | Ecological Sensitivity | Creativity | Diligence |
| . . . | . . . | . . . | . . . | . . . |

Ronald Sandler, "A Theory of Environmental Virtue," *Environmental Ethics* 28, no. 3 (2006): 243. For an updated version of this table, see Ronald Sandler, *Character and Environment* (New York: Columbia University Press, 2007), 82.

moral subject (or agent) and moral object (or patient). At the center of environmental action ethics, one finds neither environmental patients nor environmental agents, but environmental actions. Understood in terms of the basic grammar of subject-verb-object sentences, we may say that whereas environmental virtue ethics centers on the *subject*, and mainstream environmental ethics on the *object*, environmental action ethics centers rather on the *verb* (see fig. 3.1).

The starting point of environmental action ethics is the idea that certain values are either implicit in or at least strongly associated with certain types of environmental action. These values come in three main types. First, there are values present in the *subject or agent* of the action; certain valuable character traits, or virtues, may be implicit in or associated with certain types of environmental action. Second, there are values present in the *object or patient* of the action; implicit in or associated with certain types of action may be the recognition of certain values present in the object. Third, there are values present in the *relation* between the subject or agent of the action and its object or patient. In the past decade or so, these have become a major focus of debate in environmental ethics, under the rubric of "relational values."[37] This latter concept may be understood both in a broad and a narrow sense.[38] In the broad sense, the concept of relational values covers all values present in the human relation to nature, including purely instrumental ones where what is valuable is only the way in which natural objects serve human subjects, from which it follows that if another object could serve the subject just as well it could take its place.[39] In the narrow sense in which I

Environmental action ethics

$$\text{Subject} \longrightarrow \text{Verb} \longrightarrow \text{Object}$$

Environmental virtue ethics                    Mainstream environmental ethics

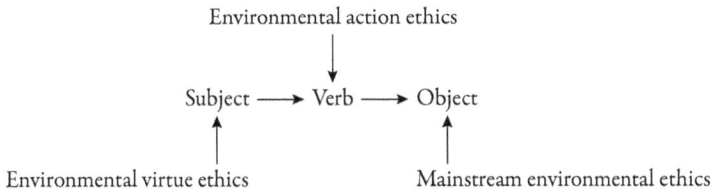

3.1 Three approaches to environmental ethics

will henceforth use the term, by contrast, the concept of relational values covers only noninstrumental relational values, where what is valuable is the relation itself, in which case the natural object involved in the relation could not in principle be replaced by some other object. Relational values in this narrow sense are of particular importance to environmental action ethics, for it is actions that concretely and practically mediate the relation between human subjects and natural objects, and, in undertaking these actions, human subjects typically both produce and make manifest the values present in their relation to nature.

Environmental action ethics, thus conceived, is not simply applied environmental ethics, not just the application of a preexisting ethical theory or stance, such as anthropocentrism or biocentrism, to some environmental action or other. Like environmental virtue ethics, the specificity of which is that it focuses on the agent, environmental action ethics is a type of environmental ethics sui generis, the specificity of which is that it focuses on the action that practically mediates the relation between agents and patients. But what is original about environmental action ethics is not simply that it is "action-centered," as opposed to agent- or patient-centered, but the fact that the actions on which it centers constitute a node around which a *nexus of values* may gather. As that which mediates the relation between human subjects and natural objects, environmental actions are in a unique position, not only to make manifest, strengthen, and produce "relational values" but also to connect and gather together certain values present in both the agent and the patient.

The existence of a nexus of implicit of associated values explains why, to each main type of environmental action, there corresponds a specific "ism." Preservationism, for example, is that position in environmental action ethics that upholds and defends the values implicit in or associated with the action of preservation. And yet, while a number of key positions within environmental action ethics—notably preservationism, conservationism, and restorationism—have been much discussed in the literature on environmental ethics, they have yet to be recognized as instantiating a specific approach to environmental ethics and so to be grouped together under a common banner. As a result, the action-centered approach to environmental ethics has tended to remain both marginalized and fragmented in comparison with the object- and subject-centered approaches that currently dominate.

Drawing on this view of environmental action ethics as an important but as yet unrecognized approach to environmental ethics, I will in what follows turn my attention to what I consider to be the four main positions that arise within it. In doing so, my aim is not only to provide a clearer understanding of environmental action ethics in general, but also to set the stage for an explicit analysis of the nexus of values implicit in or associated with that particular environmental action with which this book is primarily concerned: the imitation of nature.

### Preservationism and Conservationism

While environmental action ethics is the latest branch of environmental ethics to have acquired a name, it was the first to have come into existence. Indeed, if the origin of environmental ethics is traced back not to its establishment as an academic discipline, but rather to the first ethical debates between environmentalists, then environmental ethics emerged at the time of the seminal opposition between preservationists and conservationists in the late nineteenth and early twentieth centuries. Preservationists, like John Muir, who helped established America's first national parks, advocated a "preservation ethic" of preserving nature untouched by humans, at least other than for purposes of low-impact recreation. They were opposed by conservationists, like Gifford Pinchot, the first head of the U.S. Forest Service, whose "conservation ethic" advocated a policy of managing natural resources so as to ensure their conservation for future use.[40]

To recent and contemporary environmental ethicists reflecting on this debate, it has typically been assumed that the distinction between preservationism and conservationism ultimately reduces to a deeper philosophical opposition between anthropocentrism and nonanthropocentrism.[41] But while there are certainly links between, on the one hand, preservationism and nonanthropocentrism, and, on the other, conservationism and anthropocentrism, it would be a mistake to reduce these links to a relation of underlying identity.[42] Preservationism does characteristically include the nonanthropocentric motive of preserving nature on account of its moral standing, but it also maintains the view that preserving nature is important, even essential, to the physical and spiritual well-being of humans, most obviously because it makes possible various different forms of bodily and spiritual communion with nature (e.g., while hiking, stargazing, painting landscapes). Further, it is not hard to see that preservationists see the relation to nature as valuable in itself, for communing with nature may give rise to unique and irreplaceable feelings of wonder, attachment, love, and belonging.[43] To reduce preservationism to nonanthropocentrism is thus to overlook the various relational values implicit in or at least strongly associated with preserving nature.

Conversely, while the conservation of natural resources currently exploited by humans is for the most part undertaken for anthropocentric reasons, it could—in theory at least—also be undertaken for the benefit of other species. The proper conservation of farmland, as opposed to its exploitation to the point of desertification or soil sterility, would be of benefit to a whole host of organisms and species beyond humans, and this may provide an additional motive for conserving nature to the main one of sustainable economic use.[44] Conserving meadowland, for example, may benefit various species of wildflowers and pollinating insects that could not survive if the land were to fall out of use as pasture and revert to forest. Moreover, even when natural entities are conserved on account of their instrumental benefit to humans, they may at the same time also be valued as ends in themselves. Just as an employer can respect the second formulation of Kant's categorical imperative—"Act in such a way that you treat humanity, whether in your own person or in the person of any other, *never simply* as a means, but always *at the same time* as an end"—when hiring someone to do a job for them, an analogous principle may apply to nature, as might occur when an organic farmer employs natural predators to control pests.[45]

Likewise, it is also quite possible for conservation to be undertaken in such a way that the relation to nature that underlies it is valued in and for itself. Such relational values as partnership and cooperation, for example, may readily find their place in conservation work. A similar point applies to the way we relate to nature as that which sustains us, though in this case the intrinsic value of the relationship is even stronger, for it is a relationship on which our very existence depends. In both cases, however, the relation is not simply instrumental; there is no other partner or sustainer that could potentially take nature's place.

Analysis of preservationism and conservationism additionally shows how certain environmental virtues may be implicit in or strongly associated with certain environmental actions. Preservationism primarily involves the virtues that Sandler groups together under the headings of respect for nature, communion with nature, and environmental activism. Conservationism, by contrast, primarily involves the virtues of sustainability and environmental stewardship. Further, since it is widely recognized that virtues require cultivation, and that the cultivation of virtues often calls for action, it follows that carrying out such environmental actions as preservation and conservation may play a major role in the cultivation of environmental virtues.

### Restorationism

Returning to the question of the different positions within environmental action ethics, let us now consider the third major position: restorationism. From approximately the 1980s onward, restorationism established itself as an important alternative to conservationism and especially preservationism.[46] A key insight of restorationism is that, at a time when relatively little wild nature remains, but vast swathes of the earth have in one way or another been ecologically degraded by human activity, there is greater scope for restoring nature than for preserving it. Moreover, restoration will often be a necessary precursor to conservation, for there is little point conserving something already badly degraded.

But restorationism does not only advance quantitative arguments in favor of the increasing importance of restoration; it also makes the qualitative argument that there are important values implicit in or associated with restoration that are absent from conservationism and especially preservationism. Many of these values are relational. When working side by side to remove rubbish from a local stream, to clean up a beach, to replant a forest, or even to realize more technically complex tasks, like soil depollution or the restoration of a disused mine, humans are not simply spectators and guardians of nature, leaving it to look after itself, or even stewards and managers of natural ecosystems, but, so it is often claimed, participants in nature's own regenerative processes.[47] Related to this is the view it implies of humans as capable of making positive contributions to nature. So, whereas preservationism would appear to suppose that the only physical impact humans can have on nature is negative, and conservationism that in conserving natural resources humans can at best have a neutral impact on nature, consuming only the "interest" on the natural capital and not the natural capital itself, restorationism underlines the potentially positive impact humans can have, at least subsequent to prior acts of destruction or degradation. What is valued, then, is above all a relationship to nature based on constructive participation.[48] In keeping with this, what restorationists sometimes claim is restored is not only nature itself but also "the human cultural relationship with nature."[49]

The nexus of values implicit in or associated with the action of restoration extends also to its object. Restorationism's ethical orientation is, however, more ambivalent than in the case of preservationism and conservationism. Preservationists typically argue that implicit in restoration projects is an anthropocentric attitude, according to which nature is an infinitely malleable resource that may be destroyed and restored at will.[50] From this perspective, it matters not whether we destroy this or that natural ecosystem; we can simply move it elsewhere or recreate it later on. Restorationists sometimes retort that the act of restoration is in fact an act of restitutive justice, a righting of a past wrongs done to the beings harmed when the biotic community in question was initially destroyed or degraded, and so not anthropocentric at all.[51] From this perspective, the possibility of restoration could never be used as an argument in favor of nature's temporary destruction, for that initial destruction would still count as a wrong.

As for restorationism's relation to environmental virtue ethics, it is clear that there are certain virtues strongly associated with restoration, especially those that Sandler classes under the rubrics of communion with nature and environmental activism. Particularly in the case of community-based restoration, one finds these two sets of virtues closely intertwined, with communion with nature taking on an active and participatory dimension, and activism assuming not the traditional form of political protest, but that of restoring both nature and the human cultural relation to nature. The environmental virtues of restorationism have not, however, gone unchallenged by preservationists. They accuse restorationists of lacking the virtues of temperance and humility, and of the concomitant inability or unwillingness to temper the arrogant impulse to interfere and meddle with nature.[52]

## Imitationism

Let us now turn our attention to a fourth position within environmental action ethics, which, by analogy with preservationism, conservationism, and restorationism, may be termed "imitationism."[53] As its name suggests, imitationism is primarily concerned not with the preservation, conservation, or restoration of nature, but with its imitation. How, then, may we characterize imitationism? Just as a key insight of restorationism is that, at a time when so many natural ecosystems have been either destroyed or degraded by human activity, there is more to restore than to preserve, so a key insight of imitationism is that, at a time when artificial systems occupy so much of the earth's surface, doing unprecedented and untold ecological damage in the process, there is more work to be done imitating nature than there is preserving, conserving, or even restoring nature.

To this quantitative belief, one must also add the qualitative belief in certain values implicit in or strongly associated with imitating nature and that set imitationism apart from other positions in environmental action ethics. It is precisely here that the principle of nature as measure takes center stage. As the very existence of biomimicry testifies, the action of imitating nature, which is grounded in the principle of nature as model, is strongly associated with another principle, nature as measure, which, as we have seen above, gives biomimicry a specifically ethical dimension. Of course, one may also imitate nature without embracing the principle of nature as measure—that is to say, without thinking that an ecological standard located in the first instance in nature should be applied to human action, technological innovation included. This is, after all, something that often occurs in biomimetics and bioinspiration, for these approaches take nature as model without systematically applying nature's ecological standard to their innovations. And yet the possibility and reality of practicing imitation independently of an ecological standard clearly does not contradict the fact, incontrovertibly demonstrated by biomimicry itself, that there exists a strong association between the principle of nature as model and that of nature as measure. As in the case of theomimicry, anthropomimicry, and pantomimicry, taking a specific entity (or set of entities) as model is strongly associated with taking the same entity (or set of entities) as measure.[54]

When one further considers that a third principle, nature as mentor, underlies and permeates the principles of nature as model and nature as measure (see chapter 4), it becomes clear that there are significant relational values implicit in, or at least strongly associated with, the action of imitation. Typically, when one takes another as a mentor, what is valued is not only the knowledge and wisdom they pass on but also the very relationship one has with them, which is often formative or transformative for the mentee, in the sense that it shapes their very identity. Further, when the mentor is at the same time a measure, when they provide a standard against which right action is evaluated, the relation clearly assumes great ethical value, for, without some measure of right action, one would lose one's way in the world and be left without any basis for discriminating right from wrong.

Let us now consider how the action of imitation relates to nonrelational values, starting with any values present in natural objects themselves. Now, since all actions are undertaken with a view to an end, we may ask who or what the ultimate beneficiary of the action of imitating nature is. In response to this question, many would no doubt claim that imitationism is anthropocentric in orientation. If, as some biomimicry advocates and practitioners suppose, the principal problem humans currently face is that their technological artifacts and systems are unsustainable and thus put humans themselves in danger, then it may appear that it is only to this problem that biomimicry provides an answer. Enlightened self-interest dictates that we value nature not only as a source of raw materials but also as a source of what one might call "raw immaterials," on which humans may draw for their own benefit. From this perspective, it is in the very nature of biomimicry to instrumentalize nature as an immaterial resource, a store of knowledge from which we may benefit. This suspicion of in-built anthropocentrism may appear to be confirmed when one notes that Benyus explicates the principle of nature as model as follows: "Biomimicry is a new science that studies nature's models and then imitates or takes inspiration from these designs and processes *to solve human problems.*"[55] So, even if we admit that there may also be some noninstrumental relational values implicit in or associated with imitation, it may still be thought that imitationism fails to recognize any intrinsic value in nature from which moral duties may derive.

In response to this, it may be objected that the direct beneficiaries of imitation do not have to be human beings. Consider fishways. Whether or not they are biomimetic with respect to, say, forms and materials, as is the case regarding rock-ramp fishways that mimic the forms and materials of the natural pools and structures thanks to which fish like salmon make their way upstream, there can be little doubt that their function is taken from nature: they allow fish to pass upstream (despite the presence of artificial dams). The motivation for enabling fish to do this may of course be that it will allow them to breed successfully and thus contribute to the halieutic resources available to humans, or even just that watching fish return to their upstream spawning grounds is a pleasant aesthetic spectacle. But it is also quite possible for fishways to be integrated into artificial dams for the benefit of the fish themselves or for the benefit of the ecosystem or ecosystems in which they participate. One may, at least in theory, imitate nature for other than human ends. From this perspective, the problems humans face may in some instances only be "human problems" in the sense that it is humans who have to deal with them, but not necessarily in the anthropocentric sense that they concern humans as the ultimate beneficiaries of the problems' resolution. It follows that there is no necessary connection between imitating nature and the realization of specifically human ends, between biomimicry and anthropocentrism.

A skeptic may of course assert that the example of fishways is the exception rather than the rule, for imitating nature applies in the vast majority of cases to products and systems that only exist for the direct benefit of humans. But the simple fact that imitation *applies* in the most part to artificial products and systems that only exist for the benefit of humans does not mean that it is necessarily *undertaken* only for the benefit of humans. The fact that humans were the only directly intended beneficiaries of the products and systems that biomimicry seeks to revolutionize does not mean that they are the only directly intended beneficiaries of the biomimicry revolution. In view of this, there can be little doubt that imitationism could potentially be nonanthropocentric in orientation, though whether or not any given imitationist is anthropocentric or nonanthropocentric will depend on how exactly they interpret the principle of nature as measure, since it is by means of this principle that ethical norms guiding the human relation to nature arise. This conclusion is supported by the fact that many of those working on the philosophy of biomimicry, including, notably, Vincent Blok and Freya Mathews, have embraced a nonanthropocentric interpretation of biomimicry.[56]

Various environmental virtues are also implicit in or strongly associated with imitation. These include virtues of communion with nature, including reverence for nature's vast experience and accumulated wisdom, as well as openness and curiosity with respect to what nature may teach us. Virtues of sustainability are also closely associated with imitation. Consider, for example, Jackson's claim that "the belief that we can never do better than nature's ways may be the only source of humility for the secular mind."[57] At a time when we—or at least the secular among us—no longer show humility with respect to the standard set by God, the imitationist principle of nature as measure calls for humility toward the ecological standard set by nature.

Such claims to virtue are not, however, likely to go undisputed. Preservationists may well object that imitationism presupposes precisely the contrary belief to the one affirmed by Jackson: the arrogant and anthropocentric belief that through imitating nature we may replace it with artificial equivalents or analogues that better serve our purposes. Imitating nature, from this perspective, would be the ultimate ally of weak sustainability, which holds that the replacement of natural capital may continue indefinitely provided only that it is replaced by sustainable technological and human capital. Contrary to Jackson's profession of humility, preservationists may thus put forward a similar critique of imitationism to the one made of restorationism—namely, that the belief in our ability to improve on nature through creatively imitating it is but another instance of human arrogance.

In presenting imitationism as a focal point for critical debate, I am presenting it in a manner that is, at least in that respect, no different from how I presented the other main positions in environmental action ethics (see table 3.3). This is quite deliberate. As I see it, there may be a multitude of different positions within imitationism, many of which, on account of its relative immaturity, have yet to be explicitly theorized. Likewise, there is clearly significant scope for attacking both

TABLE 3.3 Positions in Environmental Action Ethics

|  | Preservationism | Conservationism | Restorationism | Imitationism |
|---|---|---|---|---|
| **Relational values** | Nature as irreplaceable object of love and wonder | Nature as partner and sustainer | Nature as benign whole to which we may positively contribute | Nature as measure for right action, and mentor from which we may learn |
| **Ethical orientation** | Primarily nonanthropocentric | Primarily anthropocentric | Ambivalent / disputed | Ambivalent / disputed |
| **Virtues** | Respect, communion, activism | Sustainability, stewardship | Communion, activism | Communion, sustainability |

specific positions within imitationism, as well as for attacking imitationism in general. Imitationism, on this view, is a broad position that, like preservationism, conservationism, and restorationism, already has many adherents, even if they do not explicitly label themselves as such, including those theorists and advocates of biomimicry who explicitly promote and defend certain beliefs and values they see as implicit in or associated with imitating nature, in particular the principle of nature as measure. But if imitationism contains many possible positions, some of which will inevitably be more cogent than others, it cannot be the final word on biomimetic ethics. A cogent theory of biomimetic ethics may share much in common with imitationism, though without simply identifying with it in toto.

But before I go on to spell out my own theory of biomimetic ethics, let us consider an obvious and important objection that may be made to all the positions in environmental action ethics discussed above—namely, the pluralist objection that preservation, conservation, restoration, and imitation are all important forms of environmental action, in which case it is a mistake to see any one of them as lying at the basis of an "ism," for doing so will accord that action a problematic claim to primacy and superiority with respect to other actions. This pluralist objection may well go hand in hand with the pragmatist objection that environmental goals stand a better chance of being accomplished in the absence of dogmatic conflicts between different "isms." Rather than arguing against one another, preservationists, conservationists, restorationists, and imitationists should look to work together, sometimes favoring one or the other of their respective actions, instead of insisting on the ethical primacy and superiority of one action at the expense of others. So, just as mainstream environmental ethics is increasingly populated by advocates of the pluralist view that humans, animals, living beings, and ecosystems are *all* intrinsically valuable, albeit in different ways and for different reasons, in which case representatives of the interests of each of these moral patients should work together in a pragmatic manner, prioritizing sometimes one, sometimes another center of value, depending on context, so an analogous position with respect to the four main types of environmental action carries obvious appeal.

A pragmatist may also object that, despite the above arguments regarding the greater practical significance of an action-centered approach compared to the mainstream object-centered approach, it runs counter to this practical orientation to insist that there is one form of environmental action that carries with it a primary and superior set of ethical values. To many restoration ecologists, for example, the vehement criticism of their work put forward by preservationists is hard to comprehend; in practicing ecological restoration, they are not thereby affirming its primacy and superiority over preservation, but only its appropriateness to certain contexts—those where nature has already been significantly degraded and human intervention would help it recover.

There is of course much truth in these pluralist and pragmatist objections to both mainstream environmental ethics and environmental action ethics. And yet there are also major weaknesses. Pluralism in environmental action ethics proposes no overall theoretical structure capable of generating original ethical content and that may in turn tell us what actions we ought, generally speaking, to undertake with respect to nature. It may well be true to say with pluralists that preservation, conservation, restoration, and imitation are all valuable, and with pragmatists that their relative value varies with context. But pluralism and pragmatism lack the theoretical framework required for establishing the conditions in which the greater value of one or the other might be decided.

Taking account of these strengths and weakness, I will in what follows put forward a theory of biomimetic ethics that recognizes what truth there is in pluralist and pragmatist objections to object-centered and action-centered environmental ethics, while at the same time possessing the sort of definite structure that provides both an overall ethical vision (i.e., a general theory of what it is that humans should do with respect to nature) and the means to come to definite conclusions in the face of certain basic dilemmas in environmental ethics.

## BIOMIMETIC ETHICS

### Imitation and Emulation

A distinction may be drawn between imitation and emulation. To imitate something is to take it as model. Nature, for example, may be imitated in the sense described in the previous chapter; models may be abstracted from nature and transferred over to technology. But what, then, is emulation, and how is it different from imitation?

The words "emulation" and "imitation" share the same Proto-Indo-European (PIE) root, *aim-, meaning to copy or to imitate.[58] The meanings of the words differ, however, in two important respects. First, emulation has an evaluative dimension in that it implies that the object of emulation is good or does something well, that it is successful in some respect or other. In imitating something, by contrast, there is no necessary judgment that the object of imitation is good or does something well; parodies, for example, are satirical or ironic imitations that manifest a negative judgment toward their object. Second, what is emulated is not some trait or other abstracted from an entity and then taken as model, but rather the *distinctive way of being* of the entity. One would not say that the Shinkansen "emulates" the form of the kingfisher's beak, but that it "imitates" it, for emulation applies only to distinctive ways of being, not to abstract traits, such as forms. Taken together, these two considerations allow us to define emulation as *striving to succeed or excel in the manner of another.*

It is also not hard to see that emulating involves taking the object of emulation as measure, since it provides the standard against which one's own success or excellence (or lack thereof) is measured. Just as to the principle of nature as model there corresponds the verb "imitation," to the principle of nature as measure there corresponds the verb "emulation." The principle of nature as measure does not, however, simply tell us that there is something good about nature's way of being, and thus also something worthy of emulation; it also tells us that that way of being provides a standard against which the "rightness" of our own way of being may be measured. There is, in other words, not simply an evaluative dimension to this principle, but a normative one as well. So, whereas one may or may not seek to emulate, say, a great artist, one should seek to emulate nature, for that is—at least in those contexts where the principle applies—the "right" thing to do. In keeping, however, with Benyus's claim that the measure or standard in question is ecological, it is not nature in general (i.e., physis) that one must seek to emulate, but one sole region of nature, the ecological. And since, as we have already seen, that region of nature is occupied by one single entity, Gaia, it follows also that the emulation in question has only one object: Gaia. To take nature as measure, then, is to seek to emulate Gaia, or—since emulation concerns distinctive ways of being—to *be like Gaia.*

## Being Like Gaia

We saw in chapter 1 that the specificity of Gaia's way of being is that she provides and maintains a home for her living inhabitants. The provision of a home is made possible by two conjoined processes that together constitute Gaian autopoiesis:

self-construction, through which living beings create conditions conducive to life, and self-delimitation, through which the habitable region of the earth is delimited from its uninhabitable surroundings. As for the maintenance of a home for Gaia's living inhabitants, it occurs through responding to perturbations that reduce habitability, whether through nullifying, suppressing, or reversing the ecological destructiveness of the perturbation in question.

The starting point of biomimetic ethics is that Gaia's way of being—providing and maintaining a home for her living inhabitants—is an *ultimate good*. But, one might ask, why is that? In the last analysis, the ultimate good on which all ethical systems are grounded is simply *self-evident*. There is something that simply presents itself as an ultimate good, something that we simply see as such. This ultimate good only comes to light, however, by way of contrast with its opposite. This may at first seem contradictory; how could something *self*-evident become evident through something *other than itself*? On closer inspection, there is no contradiction here. Just as self-producing beings, in producing themselves, also delimit themselves from what thereby becomes other than themselves, so the good only *stands out* as good in contradistinction to its other, to the bad. The good cannot *ek-sist* without the bad.

Virtue ethics, for example, rests on contrasting virtues, such as courage, moderation, or generosity, with vices, such as cowardice, excess, or miserliness. In a world in which vicious character traits were entirely absent, even if only as imagined possibilities, it would be meaningless to propose an ethics centered on the cultivation of virtue; virtue only exists in contrast to vice. In a similar vein, utilitarianism rests on contrasting pleasure or happiness, held to be self-evidently good, with pain or suffering, held to be self-evidently bad. But if pain and suffering did not exist, we would live in blissful ignorance of the goodness of our pleasure and happiness, and it would be meaningless to seek to avoid the former and pursue the latter. As for Kantian deontology, it likewise rests on contrasting those maxims that may consistently be universalized and are thus held to be self-evidently good (or at least not self-evidently bad), with those maxims that contradict themselves when universalized and are thus held to be self-evidently bad. But if actions based on nonuniversalizable maxims did not exist, even as imagined possibilities, if we lived in a world from which lying, breaking promises, stealing, and killing were entirely absent, we would live in blissful ignorance of the goodness of truth-telling, keeping promises, respecting the property of others, and letting others live.

In a similar vein, I suggest that the provision and maintenance of a habitable earth is something that only comes to light as self-evidently good when contrasted with its contrary: the destruction and loss of a habitable earth. But—and this is the key difference with the various ethics I have just been discussing—prior to the contemporary period, this self-evident bad had not yet come to light. Soothsayers and prophets may at times have foretold the end of the world, but the anticipated apocalypse was never ecological in nature; it was not caused by a failure to respect ecological standards, but some other form of wrongdoing, usually religious. We lived in blissful ignorance of the goodness of Gaia.[59] By contrast, as the earth becomes less and less habitable, the more it becomes self-evident that Gaia's ecological way of being, the provision and maintenance of a home for her living inhabitants, is an ultimate good. Ultimately, however, this goodness is something that one must simply *see*, even if such seeing is only possible against the alternative of its contrary; there exists no *argument* for this goodness any more than there exist arguments in utilitarianism in favor of seeing pleasure or happiness as a good. Goodness, like nature, simply presents itself.[60]

An important feature of the ultimate good described above is that what it proclaims to be an ultimate good is not an object, Gaia, but rather the *way of being* of that object—namely, providing and maintaining a habitable earth. This is important, for if it were Gaia qua object that were good, our duty would be toward that object, but if it is Gaia's way of being that is good, an alternative ethical possibility presents itself: adopting that way of being as our own. Indeed, if Gaia's way of being is an ultimate good, then it would seem to follow that, to the extent that doing so is possible, we should adopt that way of being as our own—that we should endeavor to *emulate or be like Gaia*.

An analogy may shed light on this. Saints in Christianity are not privileged *objects* of ethics, as if we had greater duties toward them than toward ordinary mortals, but moral exemplars—that is to say, beings who set a high moral standard (albeit one that could never quite measure up to the standard set by God Himself) against which other mortals may measure the goodness of their own character and actions. This analogy should not be taken to imply that Gaia is a moral or ethical being (i.e., an ethical agent or subject); the claim is only that Gaia's *ecological* way of being is an ultimate good, not that it is itself an *ethical* way of being, for an ethical way of being requires conscious awareness of how one is being. When we adopt Gaia's ecological way of being as our own, however, that is to be in a way that is ethically good, for it is to make a conscious endeavor to be good.

One could perhaps object here that, while Gaia's way of being may indeed be an ultimate good, it is not one that we should try to adopt, for Gaia is so different from us that it would be impossible or inappropriate for us to adopt her way of being as ours. After all, whereas a saint is another being of the same basic type (i.e., a human), Gaia, one might think, is a being quite different from us, in which case it would be impossible or inappropriate for us to seek to emulate her. We may, from this perspective, be better off simply admiring Gaia from afar, rather than emulating her, letting her be, rather than trying to be like her. One may even worry that to be like Gaia would be to be inauthentic, for it would be to take our way of being not from ourselves, but from another. And yet, while we are not Gaia, we do belong to Gaia, and in that case to be authentic, to be ourselves, is to act in a manner appropriate to beings that belong to Gaia.[61] Indeed, if we do not do this, if, instead of following Gaia in being ecological, our acts contradict Gaia's ecological way of being, then, precisely because Gaia's way of being is an ultimate good, we will have become a force for bad, a force internal to but in conflict with Gaia. Only by being like Gaia can we put an end to our current conflict with Gaia and instead come to participate constructively in the being and existence of the home of all that lives, ourselves included.

But what would it mean, concretely speaking, to be like Gaia? The first thing we would need to do is to participate in Gaian self-construction by helping create conditions conducive to life. This is, however, the opposite of what we are for the most part currently doing. Our impact on Gaia's living inhabitants is at present overwhelmingly negative, as evidenced by dramatic rises in the rate of species extinctions and radical declines in wildlife populations. Even Gaia's human inhabitants are increasingly threatened; as human societies develop, they are creating conditions that are not even conducive to themselves, as evidenced by such phenomena as climate change, resource depletion, pollution, and rising sea levels. The second thing we would need to do is to carry out our activities within— and potentially also participate in the production of—the existing limits or boundaries set by Gaia, notably those of the terrestrial atmosphere. An example of success in this endeavor is the Montreal protocol, which phased out the use of chlorofluorocarbons (CFCs), the result being that ozone levels have now stabilized and may even be recovering. An example of failure is our current inability to regulate and control our greenhouse gas emissions, which are increasing rapidly and leading to a potentially catastrophic decline in habitability for both humans and nonhumans alike.

As regards the maintenance of habitability, we saw earlier that this takes place in Gaia through responding to perturbations that reduce habitability. Again, this is not something we are currently doing. Instead, we are reducing habitability on a scale and at a speed that may well be unprecedented in the entire history of the planet. Through such things as habitat destruction, greenhouse gas emissions, and the pollution of air, water, and soils, we are making the earth increasingly uninhabitable. Far from being like Gaia in responding to perturbations that reduce habitability, we ourselves are the primary source of these ecologically destructive perturbations. But there is no necessity in being like this; through recognizing Gaia's way of being as good and adopting it as our own, we may ourselves respond to these anthropogenic perturbations in ways that would maintain habitability. Far from pursuing our ecologically destructive—and, ultimately, also self-destructive—path, we may emulate Gaia in seeking to suppress the ecologically destructive phenomena in question, to nullify their ecologically destructive effects, or to reverse these effects such that they become ecologically productive.

Consider the burning of fossil fuels. We may of course seek simply to suppress this phenomenon, turning instead to other energy sources, such as solar. But we could also seek to nullify its ecologically destructive effects, most obviously through employing carbon capture and storage. And, lastly, we could seek to reverse its ecological destructiveness—for example, by planting forests or farming in ways that would absorb the excess carbon dioxide, thus transforming something ecologically destructive into something ecologically productive (of forests and biologically rich soils). Consider also the case of plastics, which have a destructive impact on the habitats of many species. Again, we could simply stop producing conventional plastics, turning to natural fibers or bioplastics as alternatives. But we could also seek to nullify their ecologically destructive effects—for example, by means of chemical additives that allow them fully to biodegrade, as is the case with oxo-biodegradable plastics.[62] And, lastly, we could seek to reverse their ecological destructiveness—for example, by breeding or engineering life-forms capable of digesting them and that would henceforth come to depend on plastics, just as the multicellular life-forms that arose in the wake of the Great Oxygenation Event came to depend on oxygen.[63]

Depending on the context, one of these various ways of responding to ecologically destructive phenomena may be more appropriate, and it may in many cases even be preferable to adopt—or at least carefully experiment with—all three. But what is of fundamental ethical importance is not which of the three types of response we adopt, but rather their underlying unity, which resides in the attempt to emulate Gaia in maintaining the habitability of the earth for its living inhabitants. In so doing, we would be not only maintaining habitability but also participating in the provision of habitability. Whether we nullify, suppress, or reverse the ecological destructiveness of our acts, the consequence will not be that we no longer act, or that our acts no longer have consequences, but rather that, no longer destroying habitability, our acts will instead participate in its provision and maintenance.

To summarize, it would seem that Gaia's way of being, which is to provide and maintain a home for her living inhabitants, is an ultimate good, and, as such, provides an ecological standard against which our own way of being may be judged. In approximating to the standard set by Gaia, human actions will participate in the provision and maintenance of a habitable earth, and, as such, tend toward this ultimate good. To be good is to be like Gaia. When humans fail to do this and tend instead toward reducing the habitability of Gaia for her living inhabitants, as occurs when we take our measure for right action exclusively from a quite different source, such as humans, then our actions may be characterized as unethical. The basic maxim of biomimetic ethics, I conclude, may be stated as follows: *an action is good if it tends to participate in the provision and maintenance of a habitable earth for its living inhabitants; it is bad if it tends otherwise.*[64]

## Environmental Actions

If the basic goal of biomimetic ethics is to "be like Gaia" and thus to participate in the provision and maintenance of a habitable earth for its living inhabitants, this still leaves open the question of which specific types of environmental action will enable us to do this. With a view to answering this question, let us turn our attention to the four basic types of environmental action discussed above, starting with imitation.

We saw earlier that a basic insight of imitationism is that there is more work to be done imitating nature than there is conserving, preserving, or even restoring nature. To say, however, that there is *more work* to be done does not mean that there is *no work* to be done carrying out these other actions. Indeed, in keeping with the idea that biomimetic ethics applies not only to technological innovation but also to other fields of human action, we shall see presently that biomimetic ethics clearly requires us *also* to practice preservation, conservation, and restoration. Indeed, these environmental actions are in fact necessary if we are to respect the maxim—integral to the emulation of nature in the sense of being like Gaia—that an action is good if it tends to participate in the provision and maintenance of a habitable earth for its living inhabitants.

There can be little doubt that preservation is an important duty of the biomimetic ethics I am proposing. The reason for this is straightforward. We could not possibly claim to be following Gaia in participating in the provision and maintenance of a habitable earth, while at the same time failing to preserve the natural habitat of its nonhuman inhabitants, for that would manifestly be to reduce habitability. If it is true that Gaia's way of being is good, and that we should thus seek to take it as a measure against which we evaluate our own way of being, then preservation naturally follows.

Further, unlike in those putatively nonanthropocentric ethics that hold that in instances of conflict between humans and nature we cannot allow humans *always* to win, but must instead develop ethical standards the application of which would *sometimes* grant nature victory,[65] the biomimetic ethics theorized here would mean that, at least in situations where it is humans acting as the aggressor, *nature would always win.*[66] After all, if, when conflicts flare up between the objectives of human development and those of nature preservation, nature is only allowed to win *sometimes*, this will serve only to slow down the rate of its destruction and not bring that destruction to an end.

Consider the common case of humans wishing to destroy a portion of natural habitat for purposes of development (e.g., for mining, agriculture, or urbanization). This aim contradicts the principle of keeping the earth habitable for its living inhabitants, for the habitat of many nonhuman inhabitants would be destroyed in the process. Humans, by contrast, would not lose their habitat, but would simply have to make do with the habitat they already have and that could include making more efficient, more appropriate, and more sustainable use of it. Of course, in those instances where it is nature that is the aggressor, in the sense of spilling into and laying waste to sites of human economic activity, as occurs when crops are destroyed by pests or infectious diseases ravage our farms or cities, it is humans and their symbionts who should "win." But the idea that there are circumstances in which maintaining the habitability of the earth for humans *requires us* to destroy yet more natural habitat is, I think, implausible.[67] Ultimately, though, whether or not I am right to affirm this implausibility, the salient ethical point is that humans should always look first to inhabit differently those parts of the earth they already inhabit, such that expansion into nature would only be ethically permissible if it could be shown to be impossible for humans to continue to inhabit the earth without doing so. The current right humans accord themselves to appropriate and destroy nature would be forfeited.

A thought experiment may help us see more clearly why preservation is a duty. This is best understood by way of contrast with Richard Sylvan's "last man example"—the thought experiment that first gave rise to mainstream environmental ethics.[68] In Sylvan's example, we are invited to imagine that following some sort of global civilizational collapse only one man is left on earth. In such a scenario, Sylvan claims, anthropocentric ethics implies that this last man does nothing ethically wrong in systematically setting out to destroy the rest of life on earth; since no other humans are harmed in the process and only his preferences count for anything, the last man may do whatever he likes, regardless of the destruction wreaked on nature. But this behavior, Sylvan claims, is self-evidently unethical, in which case anthropocentrism is untenable.

The thought experiment I would like to propose is in some respects the exact opposite of Sylvan's. Instead of inviting us to imagine that there is one human left on a planet otherwise inhabited by nonhumans, it invites us to imagine that there is one nonhuman left on a planet otherwise inhabited by humans. More specifically, it invites us to imagine a world in which, thanks to the replacement of natural ecosystem services by technological substitutes, human civilization has been made sustainable in the near total absence of other life-forms. Energy would be generated directly from the sun using artificial photosynthesis, artificial food synthesized directly from water, sunlight, and inorganic nutrients, carbon dioxide sequestered by artificial trees, nutrient cycling undertaken in artificial cycles, and aesthetic delight provided by plastic plants and robot animals.

Where this example differs from Sylvan's, then, is primarily in introducing the idea that, through technology, human civilization could theoretically be sustained indefinitely despite nature's near total destruction, in which case there is no need to suppose the existence of a single *last* human indifferent to nature's destruction; technology makes it theoretically possible for the needs of *all* humans to be met even in a world from which nature has been almost entirely eradicated. Further, in the case of the last nonhuman example, the destruction of nature would be far from gratuitous. If our sole aim were to maintain the habitability of the earth for ourselves, there are rational reasons to replace nature with artifice—most obviously, increased control—for artificial entities designed in ways of our choosing are more likely than natural ones to behave as we wish.

But what does the last nonhuman example purport to show? If we had no duty to participate in the provision and maintenance of a habitable earth for humans and nonhumans alike, but only for humans alone, there would not appear to be anything ethically wrong with replacing even the very last nonhuman living being with an artificial substitute. There would be no ethical barrier, in other words, to replacing the totality of nonhuman life on earth with artificial substitutes, provided only that these substitutes function in such a way that they allow humans to meet their needs. And yet there is, I think, something self-evidently unethical about replacing the last nonhuman being with an artificial substitute, something self-evidently unethical about destroying every last living being other than humans.

That it may be viable to pursue a strategy of replacement to the point where every single nonhuman life-form is left to die, every other species driven extinct, is of course highly implausible. But the point is not that we may someday end up in a situation where only one nonhuman remains, and we are faced with the choice of whether to preserve it, but that, unless we accept the self-evident good revealed by this example, there is no ethical obstacle to our heading as close to that scenario as we arbitrarily choose.

If this thought experiment still appears somewhat extreme, consider a variant, the "last natural habitat example," in which it is not a question of destroying the last nonhuman organism, but rather of destroying the last natural habitat. In this latter example, there would be no natural habitat left, no home for other beings outside of the artificial habitats created by humans—save one. With the exception of that last natural habitat, the only fish left in the seas would be in fish farms, and the only plants and animals left on land would be in farms, plantations, gardens, parks, zoos, and other artificial habitats. In this situation, I claim that it would again be self-evidently wrong to destroy this last natural habitat, even if doing so were perfectly sustainable. And if it is not ethically permissible to destroy the last natural habitat, then the same, I claim, is true of every natural habitat prior to the last.

What these two examples purport to show is that providing and maintaining a habitable earth only for humans—or, in the last natural habitat example, for humans and their cultivated or domesticated biological symbionts—is self-evidently unethical. By their very nature as examples, they do not present an *argument* in favor of the duty of preservation, but attempt instead to show, by means of an imaginary scenario, the wrongness of its contrary (and thus also the rightness of preservation). Just as I earlier claimed that the self-evident goodness of a habitable earth becomes apparent through considering its contrary, so the goodness of maintaining habitability not only for the earth's human inhabitants but also for its nonhuman inhabitants, becomes apparent through considering its contrary—that is to say, a world in which habitability has been maintained only for humans, or, in the case of the last natural habitat example, only for humans and their cultivated or domesticated biological symbionts.

It is important to realize, however, that the primary aim of these two examples is not to provide a more relevant demonstration than Sylvan's of the wrongness of anthropocentrism, but to show the duty of a specific type of environmental action: preservation. Indeed, what I claim is shown to be self-evidently wrong in these examples is not simply the failure to treat nature as something intrinsically good, something with moral standing, but the idea that human physical and spiritual fulfilment could be attained in a world from which other living beings or natural habitats were entirely absent. It may be possible to *entertain* the idea of human beings inhabiting a world from which other living beings or natural habitats are absent, and yet in entertaining the idea its self-evident wrongness immediately becomes apparent.

The claim that failing to preserve nature is wrong is not only true, however, in the sense that it violates various important intrinsic and relational values, but also because the preservation of nature is necessary if we are to take it as model, measure, and mentor. For example, in order for Jackson to take the native Kansas prairie as providing the standard against which he measures the effectiveness, appropriateness, and sustainability of his farm, the native prairie, or at least substantial parts of it, must still be intact, and the same is true of any other attempt to take the native ecosystem as providing a measure for environmental action. If the native ecosystem is only intact in fragments and if even these are at least partially degraded, then there is clearly much less scope for looking to it as a source of ecological standards. An analogous point applies to Gaia as a whole. As Gaia's ability to provide and maintain a home for her living inhabitants decreases, as climate regulation breaks down, as the oceans heat up and acidify, as synthetic chemicals disrupt trophic chains and cycles, we may also lose the possibility of seeking measures in the various ways in which Gaia provides and maintains habitability for her living inhabitants in the first place.

Of course, it could be objected here that, in order to take nature as model, measure, and mentor, we do not require it actually to exist, for we may imitate a natural model or respect a natural measure even after the being or system from which it originated has been destroyed. One could easily imagine a world in which human beings were required to carry out a systematic scientific study of any native ecosystem prior to its destruction, precisely so that the information gathered could come to function as a future source of models and measures.

Ethical considerations aside, there are two main problems with this idea. First, any information gathered will inevitably be partial, and we may therefore miss various potential models or measures that, had the ecosystem not been destroyed, would have remained available. Second, the information gathered would be "fossilized," by which I mean that it would not only be static and unchanging but also rooted in an ever more distant past. It would tell us how nature did things in a specific place and at a specific time in the past, but it would not tell us how nature would do things in the same location today or how it would adapt to the changes that we are currently facing, including, most obviously, climate change. In short, the more we destroy nature, the more we destroy the essential source of models and measures thanks to which we might inhabit the earth in a way that allows not just us, but its nonhuman inhabitants as well, to have a home. So, far from imitation and emulation being opposed to preservation, they actually require preservation.

Another objection that would likely be made to the alleged duty of preserving nature is the one often heard today that wilderness is a myth, that there are no natural ecosystems left to preserve, for they have all been in one way or another modified by human activity and thus rendered artificial. Now, it follows from the application of the concept of physis to ecology that any sub-Gaian ecosystem is "partly artificial" inasmuch as it is delimited by us. But it is one thing to delimit a sub-Gaian ecosystem and something quite different to construct an artificial ecosystem in the manner of technē (i.e., by gathering together the four causes). It is, for example, one thing to assemble plants and animals together and to organize them into an agricultural system, and quite another simply to delimit a certain region of Gaia with a view to preserving it, thus retaining it as natural habitat. Indeed, such is the difference that we may in a qualified sense still refer to such sub-Gaian ecosystems as "natural," just as we may refer to parts of the human body as "natural," despite the fact that they are delimited by us, not by themselves. Of course, having delimited a region of Gaia and set it aside for preservation, one will often modify it—for example, by walking through it, or, more substantially, through setting up, say, the occasional toilet, café, road, or parking lot. But one does not turn a natural ecosystem into something artificial, something constructed by us, by actions such as these, the precise aim of which is only to facilitate human access to the natural ecosystem, rather than to construct it by imposing form and function on a set of biotic and abiotic materials.

As for cases where human impact is more substantial, such as when a natural ecosystem is significantly modified through, say, wiping out various top predators, as occurred when humans first arrived on and settled the American content, there is again no reason to think that this turns the ecosystem into something *constructed* by humans. After all, the arrival of other species may also modify natural ecosystems, but that does not make the ecosystem artificial. To think that human beings may make natural ecosystems artificial simply by impactful contact with them, is not only to assume, and without justification, that these impacts are essentially different from the impacts of other species, simply because they come from humans, but also to misunderstand the nature of artifacts. Only by constructing and delimiting things through the bringing together of forms, materials, generative processes, and functions do we bring artifacts into existence. And, in that case, preservation can perfectly well take place with respect to natural ecosystems that have been delimited and even substantially modified by humans, provided only that they have not been constructed by humans, as is the case in agricultural systems, for instance.

Let us now turn our attention to conservation. If, by "conservation," one means the conservation of natural capital for future use, there can be little doubt that we have a duty to conserve nature, for to run down and squander the dwindling reserves of natural capital that remain would be to put our own habitation of the earth in danger. It is also important to realize that nature has much to teach us about how we might practice conservation effectively, appropriately, and sustainably. For example, whereas industrial farming tends to deplete the soil through erosion and the sterilizing effects of synthetic chemicals, natural terrestrial ecosystems tend to accumulate fertile soil over time. To apply nature's ecological standard thus gives us the ideal of agricultural systems that do not simply conserve the natural capital of the soil, but that even allow it to accumulate. A similar point may apply to natural capital that is not directly involved in cultivation, such as wild fish stocks, which could in theory accumulate over time while also being caught for human consumption.

This is, of course, an ideal, and indeed it belongs to the very principle of nature as measure that we could never do better than nature as regards the accumulation of natural capital. But it is also true that without applying nature's ecological standard our attempts to conserve resources run the significant risk of not being simultaneously effective, appropriate, and sustainable. For example, if we see the ideal as only conserving but not accumulating natural capital, and if we then fall consistently short of that ideal, as is often the case with ideals, then in the long run our actions will be unsustainable and ecological degradation and eventually collapse will ensue.

But conservation is more than just a duty that derives from nature's instrumental value to us as "natural capital." I observed earlier that conservation characteristically realizes a valued relation to nature as both partner and sustainer. What I now wish to add is that, since we benefit from this relation, it is quite proper to show gratitude for the various "services" nature provides, and in a way that would be inappropriate in the case of tools and instruments created by humans. While one may be grateful for the existence of cars and computers, this gratitude is really only to those who invented and produced them, not to the entities themselves. In the case of self-producing nature, however, since there is no distinction between the natural entity and its producer, it is to the natural entities themselves that gratitude is owed. This relation—based on giving and gratitude for what is given—is thus fundamentally different from an instrumental relation, in which the entity is thanklessly used and exploited. From this perspective, the very conception of nature in terms of natural capital is potentially very misleading, for it envisages our relation to nature in purely instrumental terms and in so doing elides our debt of gratitude to nature.

The final type of environmental action to consider is restoration. Does the maxim according to which an action is good if it tends to participate in the provision and maintenance of a habitable earth for its living inhabitants call on us also to undertake ecological restoration? Is ecological restoration a duty? Before answering this question, let us note that, though rarely recognized as an instance of biomimicry, restoration typically takes nature as model, and in some cases also as measure. If nature is to be restored either to something like how it was, or to something like how it currently is in adjacent or similar native ecosystems, as is very often the explicit intention of ecological restorations, then nature in one way or another clearly provides the model for the restoration project.[69] Indeed, without some sort of natural model, which may of course be imitated more or less creatively, it is hard to see in what sense one could talk of restoration at all.[70] Similarly, it is not uncommon for nature, in the form of native ecosystems, to furnish ecological standards (e.g., of biodiversity, nutrient cycling, water purification) against which the success of restoration projects is measured.[71]

But the fact that one may use natural models and measures in restoration ecology does not imply that, in every instance where nature has been degraded, and restoration is thus a theoretical possibility, it is a duty. If restoration is to be undertaken, ecological standards should certainly apply, but these standards could equally well be applied to other types of environmental action undertaken in the same location, such as imitation (seen as something distinct from restoration).

With this in mind, it seems to me that, when it comes to restoration, biomimetic ethics is more ambivalent than in the cases of imitation, conservation, and preservation. In some theorizations of biocentric ethics, when a native ecosystem is destroyed an injustice is committed toward those beings that died in the process, in which case its restoration becomes a duty of restitutive justice.[72] From the perspective of biomimetic ethics, by contrast, the only relevant question when deciding whether or not to restore a portion of wild nature is, as always, whether doing so will tend to maintain the habitability of the earth for its *living* inhabitants, not for any *previous* inhabitants, even if their initial destruction was unethical. And since Gaia's living inhabitants are both human and nonhuman, it would be theoretically permissible either to restore a degraded ecosystem to something like its natural state, or to use it for the primary benefit of humans—for example, by using it for agricultural, industrial, or residential development, provided that in doing so we seek to respect nature's ecological standards.

But this ambivalence does not necessarily imply a stark dilemma between the conflicting options of restoration and imitation. It may also be seen as providing an opportunity to combine the two—for example, by constructing "buildings like trees" for human habitation, while also planting—or letting spontaneously grow—various indigenous trees that would in turn provide a habitat for other indigenous life-forms. Nevertheless, given the scope for humans to inhabit the earth much less wastefully, most obviously by means of dietary changes, and given also the fact that a successfully restored nature will be closer to Gaia's way of being and thus of greater benefit to her overall habitability, as a general rule it seems likely that restoration objectives should be prioritized, whether in the case of binary choices between restoration and imitation, or when the two are combined and the indigenous ecosystem is both restored and imitated.

*Environmental Objects*

Having considered the relation of biomimetic ethics to environmental action ethics, let us now turn our attention to its relation to mainstream environmental ethics, and thus also to the debate between anthropocentrism and nonanthropocentrism. Now, biomimetic ethics is clearly not reducible to some position or other within mainstream environmental ethics, for it centers on a specific action, imitation, and an associated way of being, emulating Gaia, and not simply on trying to identify those beings, whether humans, animals, living beings, or ecosystems, that have "moral standing" and on that basis establishing the relevant duties and obligations we have toward them.[73] And yet biomimetic ethics clearly overlaps with mainstream object-centered environmental ethics inasmuch as, in affirming our duty to participate in the provision and maintenance of a habitable earth for its living inhabitants, both human and human, its *orientation* is manifestly nonanthropocentric.

But while biomimetic ethics is opposed to anthropocentrism, this is not to say that it favors nature over humans. Unlike "ecofascism," it clearly does not imply or advocate the sacrifice of either human beings or the artificial habitats on which they depend for the benefit of the ecological whole, for human beings are inhabitants of Gaia, and it is for them, too, that the habitability of the earth is to be maintained. Further, in cases of conflict between humans and other beings, biomimetic ethics proposes a simple rule: since what is right is to participate in the provision and maintenance of the habitability of the earth for all of its living inhabitants, not just for humans, we should not only resist the temptation to colonize yet more of the earth's surface but may also defend our artificial habitats from destructive incursions from other species. Since our ethical duty is to participate in the provision and maintenance of a habitable earth for its living inhabitants, we must seek both to preserve what is left of the natural habitats of other species, whether from direct human colonization or from the invasive species we bring with us, while seeking to conserve the artificial habitats on which we depend in the face of threats to their existence from other species.

If biomimetic ethics is nonanthropocentric in orientation, then we must also consider how it relates to the two main existing branches of nonanthropocentric environmental ethics: biocentrism and ecocentrism. The basic aim of biomimetic ethics—to participate in the provision and maintenance of a habitable earth for its living inhabitants—may at first sight appear biocentric in orientation, for it would seem that the ultimate beneficiaries of pursuing this aim would be living beings in general (societies and species included). But our duty here is not directly to living beings, but to maintaining the conditions in which they may flourish. In preserving a native ecosystem, we give the beings that inhabit and compose it the opportunity to flourish, but this does not mean we have direct ethical duties to them in the same way that, at least in some circumstances, we do to other humans. One does not, for example, have a duty to help a diseased or injured plant encountered in a native ecosystem in a way that one (or, at least, someone) may have a duty to help another diseased or injured human being, but only to maintain their habitat and thus also the basic condition of their potential flourishing.

But this is not to say that biomimetic ethics is ecocentric either, at least if by ecocentrism is meant the idea that it is for the benefit of one or more ecological entities, such as Gaia, that our environmental actions should be undertaken. From the perspective of biomimetic ethics, the main weakness of ecocentrism derives from the very nature of object-centered environmental ethics: it can see in ecological entities only intrinsically valuable objects, such that it is these objects, as opposed to their way of being, that are the ultimate good. Their living inhabitants thus come to appear as having only instrumental value to the ecological whole, just as the various cells and organs of a living being are often assumed to have only instrumental value to the biological whole.

But this is to make a fundamental mistake. The nature of Gaia is to provide and maintain a habitat for her living inhabitants, not to pursue her own existence by instrumentalizing her inhabitants to that end. Likewise, what is *good* about Gaia is that she provides a home for her living inhabitants, not her continued existence as some sort of planetary superorganism, as in the case of the "biocentric earth ethic" entertained by Callicott.[74] Although Callicott is right to reject the view that our duty is toward Gaia, conceived as some sort of superorganism, he is mistaken to conclude that any "earth ethic" we might put forward can only be anthropocentric. Once Gaia's way of being is properly understood, once that way of being is understood to be good, and once we come to realize that we should take it as a standard against which we measure our own way of being, we can come to see that our duty is to participate in the provision and maintenance of a habitable earth for *all* of its inhabitants, not just for humans.[75] Just as what is good about human communities is not their continued existence as supraindividual wholes, a view that serves to legitimize the instrumentalization and sacrifice of their members, but rather that they provide the conditions in which their members may flourish, the same, I claim, is true of the global biotic community—that is to say, of Gaia. What is good about Gaia is not simply that she endures, but that she provides and maintains a home for, and thus also the conditions for the potential flourishing of, her living inhabitants.

In view of this, we can see that biomimetic ethics combines elements of both biocentrism and ecocentrism, while at the same time placing neither biological nor ecological entities at its center, but an ecological way of being: participating in the provision and maintenance of a habitable earth for its living inhabitants. So, whereas classical biocentrism and ecocentrism institute an opposition between biological individuals and ecological wholes by seeing them both as intrinsically valuable objects whose interests or well-being may come into conflict, the ethics I am proposing overcomes this opposition by holding that our primary ethical duty is neither to the ecological whole nor to the biological individuals that inhabit it, but to being in the same way as the ecological whole, Gaia, the specificity of which is that it provides and maintains a home for its living inhabitants: a place where they all—both human and nonhuman—have the opportunity to flourish.

## Environmental Virtues

Let us now consider the relation between biomimetic ethics and environmental virtue ethics. In keeping with the idea that biomimetic ethics calls for a wide range of environmental actions, while at the same time combining elements from all the main positions in mainstream object-centered environmental ethics, it also calls for the cultivation of many environmental virtues already posited and theorized within environmental virtue ethics. And yet, thanks to the original relation toward nature that it proposes, biomimetic ethics, as we will see presently, often interprets and theorizes these virtues in new and original ways.

In order to give structure to my discussion, I will draw on the typology presented by Sandler, which recognizes five types of environmental virtue: 1) sustainability; 2) communion with nature; 3) respect for nature; 4) environmental activism; and 5) environmental stewardship (see table 3.2).

The first virtue of sustainability I shall discuss, temperance, is fundamental to biomimetic ethics. The original and profound meaning of "temperance" is visible in the Latin *temperare*, which means to "observe proper measure."[76] In view of this, we can see that where biomimetic ethics innovates in its understanding of temperance is in the claim that the proper measure of environmental action

derives from nature, and more specifically from the way of being Gaia. This inno-
vative understanding of temperance in turn provides a ground for the common
but derivative understanding of temperance as moderation and self-restraint; it
is precisely with a view to emulating Gaia in providing and maintaining a habitable earth for its living inhabitants that we should observe moderation and
self-restraint.

In practice, this could perhaps involve the adoption of various relevant principles or strategies that Benyus thinks may be found in nature. These include
"nature uses only the energy it needs" in order to curb wasteful energy use; "nature
curbs excesses from within" in order to establish norms, laws, and regulations
against excessive production, consumption, and waste; and "nature taps the power
of limits" in order to help us understand the correlative benefits that may accrue
from the observation of proper limits, such as, to give just one example, the finan-
cial and health benefits of walking or cycling as opposed to driving.[77]

The second virtue of sustainability I would like to discuss, humility, follows
on from the first. To observe nature's proper measure, and thus also to observe
the virtue of temperance, is to uphold the belief that, as Jackson puts it, "we can
never do better than nature's ways," and that, as we saw above, "may be the only
source of humility for the secular mind."[78] It is also revealing to note that "humility" shares with the word "humanity" (but not with "man") the common root,
*humus*, meaning "earth."[79] To be humble is to recognize one's essential belonging
to the earth, and thus also the need to take that to which one belongs as providing the measure by which one's own way of being is judged. By contrast, to claim
for oneself or one's species the standard with respect to which others are inevitably deemed to fall short is the very definition of arrogance (from the Latin, *arrog-
are*, meaning "to claim for oneself").[80]

The final virtue of sustainability I shall discuss is attunement. To be attuned
to nature involves both listening to and harmonizing with nature. But in believing ourselves both superior to and apart from nature, the virtue of attunement
was forfeited. No longer attuned to nature's energy flows, her rhythms of growth
and decay, or the manifold specificities of her myriad places, we built a world for
ourselves—a world of placeless shopping malls, airports, hotels, factories—that was,
as it were, deaf to nature, that could no longer hear nature's song. To take nature
as measure, by contrast, would involve reattuning ourselves to nature. To run our
economy on renewable energy, as occurs in nature, would involve reattuning ourselves to natural energy flows. To produce and consume seasonally, as occurs in

nature, would involve reattuning ourselves to natural cycles and rhythms of growth and decay. And to attune ourselves to the specificities of place, as also occurs in nature, would involve reattuning the world we have built to the specificities of the local geography, climate, and wildlife.

The second set of environmental virtues—virtues of communion with nature—includes such virtues as wonder, openness, attentiveness, and appreciation. In addition to ontological wonder at being itself (physis), biomimicry adds the wonder of the biologist at the workings of living beings and the wonder of the ecologist at the way they create habitats for one another, as well as, collectively, a habitable planet. Openness toward nature takes above all the form of willingness to learn from nature, of open-mindedness to the innumerable lessons she has to teach us, and that are not limited to technical lessons about how we are to design and make things, but include also ethical lessons about how we are to dwell. As for attentiveness, in the present context it involves an ear for detail that goes beyond just general open-mindedness, a willingness to take time to study and understand the nuances and subtleties both of nature's myriad characteristics and of what we might learn from them. And, lastly, to the obvious environmental virtue of appreciating all the things nature does for us, and that together provide us with a habitat in which we may flourish, biomimicry adds appreciation of what nature may teach us, of nature's role not only as sustainer and provider but also as mentor.

The virtues of respect for nature are those environmental virtues that most directly correspond to the duties toward nature theorized by mainstream object-centered environmental ethics. The first of these virtues is considerateness. In the present context, considerateness requires us to take account of the basic ecological need of both humans and nonhumans for habitats in which they may flourish. Consider the case of endangered species. There are many species currently in danger of extinction, whether through direct habitat destruction, pollution, climate change, or hunting and fishing. To show considerateness toward these species would involve seeking to participate in the provision and maintenance of a habitat in which they may flourish, including by preserving their remaining habitat, restoring their destroyed or degraded habitat, or significantly reducing anthropogenic drivers of habitat change (e.g., pollution, greenhouse gases). For example, in cases where habitat fragmentation has created isolated communities of inbreeding organisms, thus reducing genetic diversity to the detriment of the species' ability to survive and flourish, the virtue of considerateness would dispose us to seek to reconnect these communities, most obviously through restoration projects. Considerateness, it would seem, is a virtue of particular importance to conservation biologists, whose work is vital in helping us participate in the provision and maintenance of the habitats living beings require in order to survive and flourish.

Another important virtue of respect for nature is justice. The classic definition of justice is "giving to each person their due." Stated like this, other species by definition lie outside the sphere of justice, at least if it is true that only humans are persons. But there is certainly nothing that prevents us from asking whether other species deserve to have their habitats destroyed by us, whether that is something they are due. And the obvious answer to that question is that they do not deserve that, that there is nothing fair or just about destroying the habitat of existing species so that humans may colonize ever more of the earth's surface for their own ends. That is not justice, giving each their due, but the satisfaction of the interests of the stronger, which would more properly be described as injustice. What is due to other species, then, is primarily a habitat in which they may flourish, though also—at least in some cases—gratitude for the various services they provide us.

Let us now consider the virtues of environmental activism, which Sandler tells us include such virtues as cooperation, commitment, optimism, and creativity. The basic goal of biomimetic ethics—participating in the provision and maintenance of a habitable earth for its living inhabitants—already implies the need for cooperation (and cooperativeness), for global problems like climate change are problems of collective action and, as such, require cooperation.

But cooperation would not succeed without commitment, for the basic task biomimetic ethics sets us is not something that we may quickly get out of the way before moving on to something else, or that we could hope to achieve with only sporadic efforts. Indeed, as an ethic, it is not something that could ever be achieved in the sense of being completed, but rather something to which we must remain committed, just as we must remain committed to interhuman ethics. We may one day succeed in bringing about a world that is significantly more appropriate and sustainable than the present one, but it will require ever renewed commitment both to prevent it falling back into inappropriateness and unsustainability and to bring it yet closer to the ecological standard set by Gaia.

As for optimism, biomimetic ethics shares something of this virtue with Leopold's land ethic. Leopold begins the "Land Ethic" by telling the story of Odysseus killing twelve slave girls on grounds of infidelity and disloyalty, the intention being to show that ethics may evolve, for we would no longer celebrate a man who held and killed slaves, even if they had betrayed him. In a similar vein, if I began this chapter by discussing the history of measures other than nature, it was in part to show that what we take as measure may change. Faced with a new evil—an increasingly uninhabitable earth—a new ethics may emerge. As the poet Friedrich Hölderlin once said: "Where danger is, grows the saving power also."[81]

It is also true that biomimetic ethics requires creativity. It will not be enough simply to demonstrate, protest, strike, and call for change, for creativity—especially mimetic creativity—will be required if we are to respect nature's ecological standard. Nature may provide us with the blueprint, but this is not to say that there is not also a new world for us to create.

I turn now to the virtues of environmental stewardship. Rachel Welchman, on whose work Sandler's discussion of these virtues is based, focuses on two virtues of stewardship, one of which, benevolence, she characterizes as forward-looking, and the other, loyalty, as backward-looking. Let us start by considering loyalty, which Welchman discusses as follows: "Our personal identities take shape through the emulation of and identification with the ideals, practices, values, and character of others. Loyalty is the acknowledgment of such a continuing gift. We are (or should be) loyal to those who have served as our templates—for it is by our identification with and participation in their practices and projects that we have developed the character traits and values in which we take pride."[82]

Where biomimetic ethics innovates is, of course, in adopting the view that it is not only to human beings that we must remain loyal in this sense—especially such important environmental stewards as John Muir, Aldo Leopold, or Rachel Carson—but also Gaia, who, in providing and maintaining a habitable earth for 3.8 billion years, may be considered the supreme "environmental steward." Likewise, biomimetic ethics also innovates inasmuch as the forward-looking virtue of benevolence applies not only to human beings, but also to nonhumans, for it is also for the future benefit of nonhumans that we must act as stewards.

———————

Ethics, as Aristotle points out, is concerned with praxis. This is why it is sometimes called "practical philosophy," in contrast to "theoretical philosophy," which covers metaphysics, epistemology, and the like. In the case of environmental ethics, however, there is a long-standing deficit of practical relevance. Technical debates drawing ever more subtle distinctions between different interpretations of intrinsic value in nature may have allowed environmental ethics to gain greater acceptance in analytic philosophy departments, but they are of little relevance elsewhere. This lack of practical relevance in turn explains Byran Norton's view that the discipline as a whole suffers from an "Achilles' heel"—namely, "its inability to secure a hearing among environmental decision makers."[83] Even the rise of alternative approaches has done little to overturn this deficit. Environmental virtue ethics is an important development, and yet underscoring the environmental import of, for instance, humility and moderation is of only indirect relevance to environmental decision-making. And while the recent explosion of interest in relational values has brought to light an important blind spot in environmental ethics, simply noting and emphasizing the various different noninstrumental ways in which humans value their relation to nature does not in itself provide any direct practical guidance.

Drawing on Norton's view that this lack of practical relevance may be overcome by rehabilitating the concepts of preservation and conservation, the present chapter has sought to develop this insight into a new approach to environmental ethics centered on environmental actions in general. So, whereas traditional environmental ethics is patient-centered, and environmental virtue ethics is agent-centered, environmental action ethics centers on the actions that *practically mediate* the relation between agents and patients.

It is not hard to see that this third approach offers a powerful response to the deficit of practical relevance. From the very beginning, the ethical debates that emerged organically from within the environmental movement hinged on the question of what it is we should do—preservation or conservation? Something similar is true of restoration, the ethical values of which have been much debated by both practicing restoration ecologists and their preservationist opponents, as well as, more recently, of imitation, the ethical values of which were first theorized—albeit only in a rather embryonic way—by biomimicry practitioners. Environmental philosophers may have joined all these debates, seeking to bring to them greater clarity, coherence, and rigor, but the debates themselves emerged organically out of the needs of practice and practical decision-making.

Environmental action ethics also more closely reflects the nature of "real world" environmental ethical debate. Environmental decision-makers do not seek first to determine which ethical theory (e.g., weak anthropocentrism, biocentrism) is the right one and then to apply it to the concrete decision they are facing. On the contrary, the debates involved in the decision center on an environmental action, or on a choice between different environmental actions, with the nexus of values associated with that action, or those actions, being invoked in various different ways with a view to helping make a concrete decision regarding the exact course of action to be pursued. One of the main advantages of environmental action ethics, then, is that it makes this practical reality explicit and in so doing allows for a much clearer and more comprehensive understanding both of the environmental actions available and of the complex nexus of values associated with them.

To foreground environmental action ethics is not, however, to reject ethical theory. What matters is not to favor practice over theory, but rather to formulate an ethical theory that is closely connected to practice. And this is precisely the case regarding the theory of "biomimetic ethics" I have put forward. Taking nature as model, and thereby imitating nature, is closely related to taking nature as measure, the principle I see as lying at the basis of biomimetic ethics. But this associated principle requires interpretation and therewith also theory. Drawing on Benyus's claim that the measure is "ecological," as well as on my own claim that the one true ecological entity is Gaia, I interpret the principle of nature as measure as meaning that we should measure the rightness of our actions against the ecological standard set by Gaia. The result is an ethical theory based on being like or emulating Gaia, and that in turn gives rise to the following ethical maxim: *an action is good if it tends to participate in the provision and maintenance of a habitable earth for its living inhabitants; it is bad if it tends otherwise.*

But, while the principle of nature as measure, thus interpreted, is closely associated with imitation and may also be said to accord this action greater quantitative and qualitative significance than others, this is not to say that it does not support other environmental actions too. Indeed, not only may we take nature as measure when preserving, conserving, and restoring nature, but undertaking these actions will in many cases be a *duty*—that is to say, something required if we are to participate in the provision and maintenance of a habitable earth for its living inhabitants.

Likewise, the fact that biomimetic ethics is closely connected to environmental action ethics does not mean that it bears no relation to those other approaches that center on the agents or patients of environmental action. The overall aim of being like Gaia focuses our attention not only on what we should *do* but also on how we should *be*, and thus also on the various virtues we should seek to cultivate.

Likewise, since Gaia's way of being is to provide and maintain a home for her living inhabitants, we must seek to relate to the "patients" of our environmental actions in much the same way, in which case biomimetic ethics is clearly nonanthropocentric in orientation, though it avoids the pitfall—inherent in a patient-centered approach—of having to choose between biocentrism and ecocentrism.

In view of all the above, we can see that the theory of biomimetic ethics I have put forward articulates what I suggest are the three main approaches to environmental ethics (agent-, patient-, and action-centered), while at the same time recognizing and drawing on the three different loci of value already recognized in the field: the human agent, the natural patient, and the relation between the two. Just as environmental action ethics may perhaps be said to "complete" environmental ethics, for it sets out that hitherto unrecognized approach that connects the other two while also providing the discipline with the practical relevance it has thus far been lacking, so biomimetic ethics offers a "complete" theory of environmental ethics: one with the potential to cover and provide answers to all the main debates that may occur within this field, while at the same time significantly increasing the relevance of these debates for practitioners and decision-makers.

## CHAPTER 4

*Nature as Mentor*

Biomimetic Epistemology

T he third and final basic principle of biomimicry is "nature as mentor." Nature, in biomimicry, assumes the role of a mentor or teacher
from whom we may acquire knowledge. So, just as one may base a
biomimetic technics on the principle of nature as model, and a biomimetic ethics on the principle of nature as measure, one may base a biomimetic epistemology—a theory of biomimetic knowledge, or perhaps even a biomimetic theory of
knowledge—on the principle of nature as mentor. Further, just as the revolutionary character of biomimetic technics and biomimetic ethics becomes apparent
through considering the history of models and measures other than nature, the
revolutionary character of biomimetic epistemology is revealed through considering the history of mentors other than nature.

To the principles of God as model and measure present in theomimicry, one
may add the principle of God as mentor. God, from this perspective, is the ultimate source of knowledge. As the creator of all that is, He must have possessed
an idea of all the different things He created prior to having created them. Knowledge of what things are, it follows, lies in the first instance with God, such that
the study of these things ultimately allows us to acquire insight into the ideas
underlying creation and thus into the mind of God. Something similar is true of
ethics. Knowledge of the good is located in God, and it may be revealed in a number of ways, whether in creation itself, which embodies the good, or through
other modes of divine revelation, such as scripture (e.g., the Bible) or Communion (e.g., prayer). Knowledge both of what is and what is good thus derive ultimately from God, in which case God may be said to occupy the role of "mentor"
or spiritual guide for humankind. More generally, the idea that God is omniscient,
that he sees and knows everything, means that everything that can be known is

already known by God; it thus suffices to establish some sort of communication channel with God for God to come to assume the role of universal mentor.

With the advent of modernity, the role of mentor falls to Man. Knowledge, henceforth, is in the first instance located in and sourced from Man himself. A key moment in the emergence of this "anthropomimetic" epistemology was the philosophy of Descartes, which begins with the self-knowledge of the thinking subject, with all further knowledge deriving from this human-centered starting point. Likewise, both sides in the epistemological conflict between empiricism, which emphasizes the role of the senses in the production of knowledge, and rationalism, which emphasizes the role of reason, presuppose the view of Man as mentor. Whether knowledge comes from our senses or from our ability to reason, it comes ultimately *from us*, for it is through the human faculties in question that it is generated, and more or less the same may be said of the other sources of knowledge recognized in modern epistemology: introspection, memory, and testimony. In the case of introspection, knowledge derives from observing one's own internal states of mind, in the case of memory from recalling one's past experiences, and in the case of testimony from preexisting knowledge possessed by other humans. Much the same principle applies to ethical knowledge. Whether it comes from the empirical study of human nature, as in Hume,[1] from the exercise of a pure reason thought to belong exclusively to humans, as in Kant,[2] or from ethical intuition, as in G. E. Moore, it is ultimately from Man that ethical knowledge ultimately derives.[3]

If the principle of Man as mentor may be said to characterize modern epistemology in general, it is equally true that in more recent times it has been challenged in various ways. Nietzsche famously claimed that knowledge is dependent on perspective,[4] in which case the idea that there may exist universally valid knowledge independent of our theories, values, and the like, becomes suspect.[5] Postmodernists go further. Bruno Latour, for example, claims that there are no objective truths, only "trials of strength" between different actors laying claim to knowledge, in which case "epistemology's sole function is to deny passionately that there are only trials of strength."[6] But if Latour rejects the idea of knowledge as the justified and true representation of objective reality, there is also a sense in which he extends the range of possible loci and sources of knowledge—or at least candidates for the title of knowledge—beyond humans to all things. If the world is no longer sharply divided between human subjects and natural objects, as is the case in what Latour calls the "modern Constitution," but consists rather of an actor-network composed of both human and nonhuman agents (or actants), the possibility arises that knowledge—or its nonmodern equivalent—may also be present among nonhumans.[7]

When scientific hypotheses are tested in the form of experiments, for instance, Latour tells us that the objects involved in these experiments may "object," as if a possible item of knowledge proposed by a human agent were capable of being disputed by a nonhuman actant, or perhaps rather by an "alliance" of nonhuman actants, such as a system or network of interacting particles.[8] To the extent that knowledge may still be said to exist in the actor network, it is distributed across all the different actors that populate the network, in which case it is theoretically possible to locate and source knowledge from any part of that network, and not just from human beings. To the principles of anything and everything as model and measure characteristic of pantomimicry, one may add the principle of anything and everything as mentor.

With the advent of biomimicry, it is neither God, Man, nor anything and everything, that assumes the role of mentor, but nature. This shift raises a number of important philosophical questions. A first set concerns the relation between biomimetic and other epistemologies. How exactly is biomimetic epistemology different from conventional epistemology, and in what exactly does its revolutionary character consist? And how does biomimetic epistemology relate to the various alternative epistemologies that have emerged in recent times, including naturalized, externalist, social, feminist, and environmental epistemology, all of which also see themselves as in some sense revolutionary? Is it just another alternative epistemology to be considered alongside the others, or is it more than that?

A second set of questions relates to the nature of biomimetic knowledge. If we may learn from nature, it would seem to follow that nature already possesses knowledge, which may thereafter be acquired by us. But what does it mean to say that nature possesses knowledge, given that many natural entities are not conscious subjects in the way that human beings are? Likewise, how are we to understand the way that nature imparts its knowledge to us, and thus also the very idea of nature as mentor? And is the knowledge we in one way or another acquire from nature all of the same variety, or does it come in multiple different types?

A third set relates to science. If conventional epistemology underpins conventional science, does that mean that biomimetic epistemology may come to underpin what we might call "biomimetic science"? And, if so, how can we best characterize conventional science, and how exactly is biomimetic science different?

## BIOMIMETIC AND OTHER EPISTEMOLOGIES

### Biomimetic and Naturalized Epistemology

Conventional epistemology, which first truly established itself at the dawn of modernity around the time of Descartes, holds that knowledge originates with and resides in human subjects. The only sources of knowledge from this perspective are the perceptual, mental, and communicative faculties of human beings: perception, reason, introspection, memory, and testimony. To the extent that nature has anything to do with knowledge, it is only as an *object* of knowledge, something that knowledge is *about*; knowledge, which originates in humans, takes nature as its object. The basic epistemological relation between humans and nature characteristic of conventional epistemology may thus be summarized as follows: *human subjects produce knowledge about natural objects.*

In contrast to this, biomimetic epistemology considers nature not as the *object* of knowledge (that which knowledge is *about*), but rather as the *source* of knowledge (that which knowledge is *from*). Humans, viewed in this manner, are not the initial producers of knowledge, who, as such, can only ever learn from themselves (man as mentor), but attentive listeners to a knowledge first produced by nature (nature as mentor). But instead of turning immediately to a direct investigation of biomimetic epistemology, let us begin with an indirect approach, comparing and contrasting it with various existing alternative epistemologies. After all, if biomimetic epistemology is indeed revolutionary in character, we must ascertain whether it is revolutionary only with respect to conventional epistemology or also with respect to these alternative epistemologies.

The first alternative epistemology I will consider is naturalized epistemology. Put forward notably by W. V. O. Quine, naturalized epistemology seeks to turn the attention of epistemologists away from the normative issue of how we ought to go about producing knowledge and toward how we actually go about producing knowledge.[9] Knowledge production, from this perspective, is a natural process occurring in human subjects and should be studied accordingly, with great emphasis placed on what the natural sciences—including psychology, neuroscience, and cognitive science—can tell us about how it occurs.

In one respect, this naturalistic approach to epistemology is similar to that of biomimicry; knowledge production is treated as a natural process. Nevertheless, in another respect, naturalized epistemology remains conventional. A general, if not necessarily explicit, assumption of naturalized epistemology is that the process of knowledge production takes place in human beings, notably via the faculties of sense perception, reason, memory, and introspection. Naturalized epistemology may open the door to drawing more directly on the observable behavior of human beings or the operations of the brain responsible for knowledge production, as opposed to the ideal or idealized processes of knowledge production theorized by conventional epistemology, but it is still within the human being that knowledge is produced. The human being may, in keeping with the identity between being and nature constitutive of naturalist ontology, be seen as a natural entity, but it is still generally assumed that it is within the human being—and, more specifically, within the human mind—that knowledge is produced. Similarly, the archetypal form of knowledge considered in naturalized epistemology is still knowledge *about* nature, such as knowledge about water's chemical composition.

There is, however, more to the relation between naturalized epistemology and biomimetic epistemology than simply a set of similarities and differences. In naturalizing processes of knowledge production and acquisition in humans, naturalized epistemology also opens up the possibility of taking these processes as models for technological innovation. By studying the cognitive processes by which humans produce and acquire knowledge, it may be possible to imitate those processes artificially, most obviously in the fields of robotics and AI. Of course, many of the epistemic functions of human cognition (e.g., calculation, reasoning, translation, voice recognition, facial recognition, manipulation of objects) have already been reproduced in various different technologies, in which case many existing approaches to robotics and AI not yet recognized as biomimetic could potentially be reconceptualized as biomimetic with respect to their functions. But there is also great scope for taking human cognition—conceived as a natural phenomenon—as a source of models for instances of artificial cognition that more closely approximate to human cognition in terms of their forms, materials, and generative processes. Machine translation, for example, could seek not only to produce a text that conveys the meaning of the original in another language but it could also study how human translators actually go about translating and then take the generative processes thus observed as models.

One could perhaps object here that the cases discussed above concern not biomimicry but anthropomimicry, for it is the human being that is taken as model. There are two obvious responses to this objection. The first is that, since human cognition is reconceptualized as a natural phenomenon, imitating the human may be seen as a specific instance of biomimicry. The second is that, despite the general assumption of naturalized epistemology that knowledge is produced only in and by human beings, there is no reason why cognitive processes responsible for the production of knowledge in nonhumans could not provide models for technological products or solutions as well. The various different ways that nonhumans produce and manipulate knowledge (e.g., animal cognition, swarm intelligence, evolutionary learning) could thus provide biomimetic models for artificial processes of knowledge production and manipulation. Nevertheless, it is true that these are *applications*, rather than *direct implications*, of naturalized epistemology, and, as such, only come to light when the naturalization of epistemic processes—whether human or nonhuman—is combined with the biomimetic principle of nature as mentor. In itself, however, naturalized epistemology remains conventional, for it typically retains the conventional view of humans as producers of knowledge, with nonhuman nature being but an object of that knowledge.

*Biomimetic and Externalist Epistemology*

Externalist epistemology comes in two forms. According to the first of these, content externalism, whether or not something counts as knowledge depends not only on facts internal to the human subject but also on external facts out there in the world that the human subject may not have internalized. It may be the case that everything to which a human subject has access justifiably leads them to conclude that they think they know something or other about the external world around them, but, according to content externalism, if the external world is not in fact how they think it is, then they do not possess knowledge about it. If, for example, the chemical composition of water was not $H_2O$, despite all available evidence to the contrary (e.g., because we were being systematically manipulated by a malicious demon), we would not in fact be in possession of knowledge of its chemical composition.

Content externalism overlaps with biomimetic epistemology to the extent that it emphasizes the role of factors external to human beings in the production of knowledge. Knowledge, from an externalist perspective, is not produced solely by internal processes taking place within humans, but in part by external nature. Nevertheless, two major differences with biomimetic epistemology remain. The first difference is that the knowledge with which content externalism is concerned is typically assumed to be *about* nature—the chemical composition of water, for example. The second is that, while content externalism holds that external nature may *participate* in the production of knowledge, there is no suggestion that external nature may produce knowledge in a way that is *wholly* external to human beings. Nature may participate in the production of knowledge possessed by humans, but ultimately that knowledge belongs exclusively to human beings. It is we, not water, who know that its chemical composition is $H_2O$. In view of these two points, it is clear that content externalism upholds the basic tenet of conventional epistemology: *human subjects produce knowledge about natural objects.*

Vehicle externalism, the second form of externalist epistemology, is of greater interest to biomimicry. If, as its name suggests, content externalism focuses on the content of knowledge, such as knowledge about the chemical composition of water, vehicle externalism focuses on the vehicle in which that knowledge is embedded. According to vehicle *internalism*, it is in the human being, whether their mind, their body (especially the brain), or some combination of the two, that knowledge is embedded. Vehicle *externalism*, by contrast, maintains that knowledge may be at least partly embedded in external entities. Mark Rowlands gives the example of a system used by Peruvian *kvinu* officers for storing information and knowledge in the form of knots—the principal advantage of the system being that one only has to remember the code necessary for deciphering the knots and not the information or knowledge they contain.[10] Other obvious examples of external vehicles include books, computers, and maps, all of which store information and knowledge in ways we may readily access, providing we know how to use and read them, the only difference with internal memory systems being that they are located outside the body. It is thus, Rowlands argues, only our traditional "internalist prejudice" that makes us see such external memory aids as not participating directly in cognitive and epistemic processes, and thus as external to our mind.[11]

The idea that cognitive tasks may be "offloaded" onto the external environment clearly overlaps with the view of biomimetic epistemology that it is in many cases preferable to draw on knowledge already present in nature rather than to attempt to create new knowledge from scratch. Indeed, from the specific point of view of biomimetic epistemology, nature—or at least those natural beings in which legein occurs—appears as a vehicle for the production and storage of knowledge. After 3.8 billion years of evolution, vast amounts of knowledge have accumulated in the natural environment, and so, rather than treating it as a purely material resource, we should come to see it as an immaterial resource as well, a vast repository of knowledge and wisdom.

There are, however, two significant differences between biomimetic epistemology and vehicle externalism. First, the emphasis of vehicle externalism has for the most part been on the offloading of epistemic processes (e.g., memory, calculation) onto *artificial* entities, rather than onto *natural* ones. The second difference concerns the basic source of knowledge in these two contrasting epistemologies. In vehicle externalism, the external objects are not *basic* sources of knowledge, but, as the very notion of *vehicle* externalism suggests, mere carriers of knowledge whose ultimate source lies elsewhere. In the case of the technological objects on which vehicle externalism tends to focus, it is humans who are the basic source of the knowledge these objects contain, for it is we who have put our knowledge into knots, books, computers, maps, and the like. Even when these artifacts participate not in the storage of knowledge, but in its production, as in the case when computers carry out complex calculations for us, it is knowledge originating in humans that has given them this capacity.

This bias toward knowledge originating in humans but transferred to technology is reflected in an idea Mark Rowlands takes from Andy Clark: "We make the world around us smart so we don't have to be."[12] In the case of biomimetic epistemology, however, it is clearly not the case that the world around us is only smart because we make it so; the world around us is already smart, and we may become smarter through learning from it. In view of this, it is clear that the overall epistemological relation between humans and nature that emerges from vehicle externalism is still the conventional one of human subjects producing knowledge about nature, the only difference with conventional epistemology being that some of the cognitive tasks involved in producing and storing knowledge are offloaded onto external entities, usually technologies specifically designed for that purpose.

## Biomimetic and Social Epistemology

At first sight, social epistemology may seem to be perfectly conventional. Indeed, it could be viewed simply as that branch of conventional epistemology that is concerned with the acquisition of knowledge through processes of social transmission, especially testimony. But social epistemology in fact challenges conventional epistemology in two main ways. First, it emphasizes the fact that the production of knowledge is very often a social achievement—that is to say, that it arises through cooperation between individuals. Second, it emphasizes the overwhelming importance of social processes of knowledge transmission, including testimony. Almost everyone knows, for instance, that the chemical composition of water is $H_2O$, but there is probably no one alive today who originally acquired that knowledge through empirical investigation, and very few who have enough experience of chemistry to be able to provide a scientific demonstration of what everyone today accepts as knowledge about water's chemical composition.

The focus of social epistemology on the acquisition of knowledge from sources outside the self clearly constitutes a significant point of overlap between social and biomimetic epistemology. One could even seek to integrate biomimetic epistemology into a modified and expanded form of social epistemology. Just as in environmental ethics one finds the idea that we must enlarge the *moral community* to include nonhuman nature, one could likewise argue that we must enlarge the *epistemic community* to include nonhuman nature.[13] The principal complication with this idea, however, is that in theorizing this expanded epistemic community we must take account of the radical epistemic differences and asymmetries between humans and nature.

A first difference concerns the way knowledge is produced in nature. The knowledge embedded in nature of how to generate usable energy from the sun is not produced in the same way as human knowledge, for it is clearly not derived initially from the senses, reason, memory, or introspection. Similarly, knowledge transmission between humans often works in ways that are quite different from knowledge transmission between nature and humans. Plants do not come forth and tell us how they generate energy from the sun in the way that engineering lecturers may tell their how students how photovoltaic panels carry out this same function. Indeed, the key process of knowledge transmission emphasized in social epistemology—testimony—does not occur at all between nature and humans.

A second difference is that, whereas in human epistemic communities the transmission of knowledge is, at least potentially, two-way, when it comes to epistemic relations between nature and humans, knowledge transmission is one-way. Humans may learn from nature, but nature does not learn from humans.[14] So, just as a potential for reciprocity is generally thought to be a key feature of human moral communities, but not of the expanded moral communities considered in environmental ethics, something similar is true of the expanded epistemic communities under consideration here.

As it stands, however, social epistemology is concerned only with interhuman epistemic relations. This allows us to pinpoint two basic differences from biomimetic epistemology, which together mean that social epistemology ultimately remains conventional. The first is that social epistemology has assumed that *other humans* are the exclusive producers, possessors, and transmitters of any knowledge that is socially acquired. That nonhuman nature may produce, possess, and transmit knowledge, even if not consciously or deliberately, has—at least as far as I can tell—been overlooked by social epistemology. The second difference is that the knowledge acquired from outside sources is typically assumed to be knowledge *about* reality or nature, as is the case, for example, in conventional scientific knowledge. This in turn explains why social epistemologists attribute such great importance to testimony—a mode of knowledge that, as we have already seen, is inappropriate in the case of biomimicry; not only does nature not transmit knowledge either intentionally or deliberately but this transmission is also not of representational knowledge about objective reality or nature. So, despite its emphasis on sourcing knowledge from others, social epistemology still tacitly assumes the conventional epistemological tenet: *human subjects produce knowledge about nature.*

## Biomimetic and Feminist Epistemology

The key insight of feminist epistemology is that knowledge is situated, in particular with respect to gender, such that men and women often have different ways of knowing, different cognitive styles, privilege different types and objects of knowledge, and so on and so forth.[15]

The idea that knowledge is situated has led many feminists to focus on its situatedness with respect to the body, and, going further, to the idea of knowledge as bodily or embodied.[16] So, whereas Descartes proposed a radical ontological divide between two fundamental substances, mind and body, locating knowledge exclusively on the side of the mind, feminist epistemology has embraced the possibility of overcoming this dualism by seeing knowledge as bodily or embodied. Consider, for example, the difference between dancing and studying mathematical physics. In the case of dancing, the knowledge involved is clearly bodily or embodied in the sense that it is produced in large part through bodily move-ment and is to a large extent embedded within the body, most obviously in the form of muscle memory.[17] In the case of mathematical physics, by contrast, the body becomes but a support—one that often goes unnoticed—for processes that appear to be entirely intellectual. The mathematical physicist may use their eyes to read, their hands to type, their lungs to breathe, and their brain to think, but these processes all fade into the background as the focus of their activity turns to the intellectual content of the subject matter. This is not to say that the knowledge of the mathematical physicist is not, from this feminist perspective, also embodied, but simply that it is much more plausible, when this sort of knowledge is taken as paradigmatic, to posit an ontological duality between mind and body and then to see knowledge as residing solely on the side of the mind.

In a similar vein, feminist epistemology has highlighted the epistemic impor-tance of listening to the body, including to bodily sensations, feelings, and emo-tions, while conventional epistemology has typically assumed that the influence of the body will lead only to problematic bias or distortion with respect to the purely intellectual processes, such as disinterested observation, reasoning, and cal-culation, by which genuine knowledge is produced.[18] So, whereas the moral epis-temology of Kant holds that pure reason alone leads to moral knowledge, which passion and the sentiments risk distorting, feminist philosophies of care under-line the positive epistemic role played by such feelings as sympathy and compassion.

There are two ways in which the epistemic importance attributed to the body in feminism overlaps with biomimetic epistemology. The first concerns the idea that knowledge may be located outside the conscious mind. If knowledge is in the first instance bodily or embodied, rather than consciously understood, then this may well coincide with—and, indeed, help us make sense of—the idea that knowledge is embedded in nature prior to being consciously appropriated by human beings.

The second overlap concerns a parallel or analogy between the idea of listen-ing to the body in feminist epistemology and the idea of listening to nature in biomimetic epistemology. Rejecting the conventional views of the body and of nature as objects to be observed and controlled, feminist and biomimetic episte-mology put forward the idea of the body and of nature, respectively, as sources of knowledge to be listened to and respected for what they may teach us.

At the same time, the manifestly partial nature of these overlaps points to a fundamental difference between feminist and biomimetic epistemology. In feminist epistemology, the bodily or embodied knowledge in question typically remains that of the human being, not of nature. Feminist epistemology may provide an alternative to androcentrism, but not to anthropocentrism. Likewise, to the extent that it is ultimately the human being, albeit in its bodily aspects, to which one listens, feminist epistemology may be said still to subscribe to the principle of man—or, perhaps, rather, woman—as mentor.

In response to this, it could potentially be affirmed that these discrepancies between biomimetic and feminist epistemology could be overcome within ecofeminist epistemology. For ecofeminists, the problematic parallels described above between the mind-body relation and the human-nature relation both derive ultimately from androcentrism. Biomimetic epistemology, from this perspective, could be seen as a distinctly feminist alternative to the conventional epistemological relation between humanity and nature; no longer seen as an object to be studied, manipulated, and controlled, nature would instead come to be seen as something to which we may listen, a source of knowledge and wisdom worthy of attention and respect. It is perhaps also significant in this context to note that biomimetic ethics places at its center the notion of emulating Gaia, a goddess, and thus, it may be thought, of being in a distinctively feminine way. To be like nature, from this perspective, would involve a shift from a masculine attitude of domination and control of nature to a feminine one of participating in the provision and maintenance of a home for all that lives. Likewise, it may perhaps be relevant to note that the first significant formulation and expression of biomimicry—and, in particular, the explicit introduction of the principle of nature as mentor—was put forward by a woman, Janine Benyus.

While there is certainly scope for further exploring the idea that biomimetic epistemology may be interpreted as ecofeminist, it is also important to realize that there may perhaps be something stereotypically masculine about the processes of abstraction and transfer involved in the principle of nature as model underlying biomimetic technics. Feminist epistemology may perhaps help us think about the concrete knowledge embedded and embodied in nature, especially the ecological way of being and knowing characteristic of Gaia, and it may also support the general idea of being attentive to and listening to nature. But the processes involved in biomimetic innovation arguably involve stereotypically masculine ways of knowing: cutting nature up, abstracting knowledge, and transferring that knowledge into new technologies or designs. From this perspective, rather than seeing biomimetic epistemology in a binary way, as an expression of distinctly feminist ways of relating to nature, it may be better seen as involving a complex articulation of both stereotypically feminine and stereotypically masculine ways of knowing.

### Biomimetic and Environmental Epistemology

The fifth and final alternative epistemology I shall consider is environmental epistemology. As formulated by Christopher Preston, the key claim of environmental epistemology is that knowledge is situated relative to place.[19]

This relation to place, Preston claims, has both a biological and a cultural dimension. Regarding the former, Preston claims that our cognitive and epistemic faculties, including perception and technical skill, have been shaped largely by the environment in which we evolved. He gives a simple example: were we to get around on all fours with our noses closer to the ground, we may well have retained a much better sense of smell.[20] But, beyond the biological heritage of cognitive and epistemic faculties embodied in our sense organs, brains, and hands, Preston thinks that there is also a strong cultural link between place and knowledge.[21] The knowledge of a given culture is at least partly determined by the geography, ecology, and climate of the region in which that culture developed.

The idea that knowledge is relative to place provides a first point of overlap with biomimetic epistemology, for the knowledge present in nature is also relative to place. Nevertheless, it is important to realize that places exist at many different levels—or, as I put it in the previous chapter, that there are "places within places." Consider the case of energy generation. Benyus at one point expresses the generalization that "Nature runs on sunlight."[22] This is not to say that the knowledge embedded in nature of how to generate usable energy from the sun is not relative to place; it is clearly relative to the fact that life evolved on a planet bathed in sunlight. Were life to evolve on, say, a highly radioactive planet orbiting a neutron star, it would likely run on the nuclear radiation internal to the planet.

But energy production is also relative to lower-level places. Certain basic features of photosynthesis (e.g., fixing energy from sunlight, carbohydrate synthesis) are common to all plants, with the minor exception of parasitic plants that obtain energy from fungi. But, whereas most plants use $C_3$ photosynthesis, some plants adapted to tropical conditions use C4 photosynthesis, which, thanks to an alternative mechanism for carbon fixation, is approximately 50 percent more efficient. Yet other plants, like cacti, use CAM (crassulacean acid metabolism) photosynthesis, which allows them to open their stomata only at night, thus massively reducing water loss when compared to C3 and C4 plants, which open their stomata to absorb $CO_2$ during the day. Likewise, even within a specific ecosystem there will be major differences in the way photosynthesis is deployed. In tropical rainforests, for example, the canopy absorbs those wavelengths of light that offer the highest-quality energy. Other wavelengths will then pass through the canopy, where they may be absorbed by lower-level leaves that have specialized in exploiting the lower-quality energy they contain.

That the knowledge embedded in nature is relative to place implies that whatever knowledge we acquire or derive from nature shall also be relative to place. If we transfer the idea that "nature runs on sunlight" into human technology, we get the idea of a terrestrial civilization whose technologies would also run on sunlight, rather than on, say, fossil fuels or nuclear energy. But, just as in nature the way that sunlight is used is quite different depending on place, so the same could be true of how our technologies use sunlight. To give just one example, it would be possible to develop translucent solar cells adapted to the exploitation of different wavelengths of sunlight and that could be superimposed on each other, just like leaves in a forest.

There are, however, two significant differences between the place-based environmental epistemology theorized by Preston and biomimetic epistemology. The first difference is simply that the only place-based knowledge to which Preston gives serious consideration is that of humans. The second is that there is no clear distinction in Preston's epistemology between learning *about* and learning *from*. This is not to say that in Preston's epistemology, all knowledge is about nature qua object; much cultural knowledge, such as that present in culinary traditions, is relative to place, but without being knowledge about nature qua object. Nevertheless, the idea that we may learn *from* nature is never explicitly theorized.

Further, the policy implications of Preston's place-based epistemology are conservative, providing only, as he himself puts it, "*additional* support for existing environmental policy rather than . . . a radical new proposal."[23] In emphasizing the situatedness of knowledge with respect to place, what Preston emphasizes is a "sense of place," and this in turn, he thinks, may give rise to the desire to protect the places in question.[24] By contrast, the idea that we may learn from the place-based knowledge embedded in nature has truly radical policy implications, including modeling our agriculture, industry and cities on—while measuring them against—indigenous natural ecosystems.

Another variation of environmental epistemology is provided by Jim Cheney, who draws on both postmodern and indigenous epistemologies.[25] Developing the well-known postmodern idea that nonhumans possess agency, Cheney argues that this agency is partly "epistemic" and that we should therefore view nonhumans as "co-participants" in the construction of knowledge,[26] as he thinks is already the case in indigenous epistemologies, which, instead of seeing nature as simply the *object* of human knowledge, conceive instead of "a relationship between more-than-human teachers and humans who have prepared themselves spiritually for the reception of knowledge and power."[27]

While there are substantial overlaps here with biomimetic epistemology, there are differences as well. A first difference is that, whereas biomimetic epistemology sees the flow of knowledge as one-way, from nature to humans, Cheney appears to see the relationship as two-way, or, in his words, as one of "reciprocity."[28] In keeping with this, his overall focus is not so much on "learning from" nature as on "knowing with" nature.[29] A second difference concerns the types of entity from which we may acquire knowledge. In alignment with his ethic of "universal considerability," according to which any being whatsoever should be given moral consideration, when Cheney turns his attention to epistemology, he appears to be of the view that we may potentially enter into an epistemic relationship with any entity whatsoever. In focusing on rocks, his thinking is clearly that if we can have epistemic relationships with these inanimate physical entities, epistemic relationships are possible a fortiori with all other natural entities, such as plants, animals, forests, and rivers.[30] But if, as I will argue later, it is only in entities in which legein occurs that knowledge may be produced, then the idea that there may be knowledge embedded in rocks requires a significant qualification: it is only to the extent that rocks are part of Gaia that we may learn from them. There may, for example, be a great deal we may learn from how certain rocks are formed, the way they absorb atmospheric gases, or the processes by which they are broken down and release nutrients for plants. But it is only from this specifically ecological perspective that we may learn from them.[31]

I would further suggest that these two differences derive from two quite different underlying philosophies: pantomimicry and biomimicry. In pantomimicry, the underlying philosophy of postmodernism, everything and anything may play the role of mentor; rocks may be mentors to us, and we may be mentors to them.

But this indiscriminate expansion of the epistemic community fails not only to recognize the important differences and asymmetries between nature and humans alluded to above but also to offer a viable theory of nature that is capable of supporting the idea of entering into some sort of meaningful epistemic relationship with it.

A third difference concerns the way Cheney draws on indigenous epistemology. Despite the manifest overlap between biomimetic and indigenous epistemology, there is a significant difference between the two. In biomimetic epistemology, the natural sciences—understood primarily as techniques for cutting nature up (e.g., into molecules, organs, flows)—are essential to the various technical and ethical lessons nature has to teach us. But, perhaps because of the close association Cheney sees between Western science and the modern epistemology of objectification and control, his discussion of indigenous epistemology does not seek to articulate it with the natural sciences.[32] This in turn means that the epistemic relationships he discusses lack much of the practical relevance of biomimetic epistemology, for they provide little or no guidance as to how we might learn from nature how to develop either new technologies or a new ethic informed by ecological science.

This is not to say that indigenous epistemologies lack relevance. On the contrary, they may hold within them important insights that could significantly enrich or complement biomimetic epistemology. More specifically, it is quite possible that they might: 1) provide important insights into the idea of learning from nature that are lacking from Western theorizations of biomimicry, including the present one; 2) enrich or expand the various different ways we might learn from nature and the different types of knowledge we might glean from nature; or 3) hold within them specific epistemic elements (e.g., items of knowledge, nuggets of wisdom) gleaned from nature and that are absent from Western thought. Moreover, there is clearly much scope for articulating these indigenous epistemologies with the natural sciences, as has been advocated by Robin Wall Kimmerer.[33] So, whereas Cheney attempts an articulation of indigenous and postmodern epistemologies that to a large extent bypasses the natural sciences, a promising path for future research would be one that articulated indigenous and biomimetic epistemologies, and in a way that accorded a much greater role to the natural sciences.

Let us now briefly attempt to synthesize what has been learned through comparing and contrasting biomimetic epistemology with both conventional epistemology and various alternative epistemologies. Without going over all the similarities and differences discussed above, it would seem that biomimetic epistemology may be said both to assimilate and to synthesize many of the key insights of these alternative epistemologies by means of an entirely new principle—nature as mentor—not explicitly present in any of them (the only exception being Cheney's environmental epistemology, which imports the idea from indigenous epistemology). Knowing is in some sense naturalized, externalized, socialized, feminized, and environmentalized, while at the same time being genuinely revolutionized; unlike in these alternative but for the most part conventional epistemologies, it is viewed not as something produced and possessed by humans, but rather as something produced and possessed by nature and that comes to be acquired by humans only later.

However, while this exercise of compare and contrast provides a preliminary indication of the nature of biomimetic epistemology, it is no substitute for the endeavor that I shall now undertake: a direct characterization of biomimetic epistemology.

## BIOMIMETIC EPISTEMOLOGY

### Natural Knowledge

Each branch of the new philosophy put forward in this book has been grounded in a creative renewal of a foundational word of ancient Greek thinking: *physis* for ontology; *technē* for technics, and *ethos* for ethics. This chapter is no exception. Drawing on the argument put forward in chapter 1 that the distinctive feature of biological and ecological instances of autopoiesis is *logos*, understood as selective bringing together, I will now show how this view of logos also makes possible a renewed understanding of *epistēmē*, and therewith also a specifically biomimetic epistemology. Key to this new epistemology is the idea that knowledge is not something separate from nature, existing only in the minds of human beings, but rather that it is present in nature and that there is thus what I propose to call "natural knowledge."

If we take human knowledge, and especially knowledge of the form "S knows that P," where "S" designates a subject and "P" a proposition (e.g., "I know that the chemical composition of water is $H_2O$"), as the paradigm of all knowledge, then it would seem that there is no such thing as natural knowledge. The process of photosynthesis in plants may involve the splitting of water into hydrogen and oxygen, but this does not imply that plants know that the chemical composition of water is $H_2O$. We cannot conclude from this, however, that nature does not possess any knowledge, but only that whatever knowledge it does possess is not of the form "S knows that P." So what is the nature of the knowledge embedded in nature? In what follows, I will argue that there are three positive ways of characterizing it: as monistic, autistic, and holistic. I will also argue that each of these three characterizations of natural knowledge corresponds to a different aspect of an idea common to them all: that *knowing in nature is one.*

Let us start by examining the claim that the knowledge embedded in nature is monistic. Monism is the thesis that being is one, especially the thesis that being and thinking are one. But what does this mean? According to Maturana and Varela's theory of autopoiesis, the organization of the living is at the same time its way of cognizing. The way of being of autopoietic beings, it follows, is also their way of thinking or cognizing; physis and logos are one. What, then, is the relation between logos and epistēmē? Today we typically believe that the knowledge possessed by human beings is produced by logos, in the sense of reasoning, for it is only by giving reasons that our beliefs may be justified and thus attain the status of knowledge. Something analogous, I claim, is true in the case of physis: knowledge is produced by legein—that is to say, by processes of selective bringing together. Natural selection, for example, produces vast quantities of knowledge. And since physis and logos, as we have just seen, are one, it follows that physis and epistēmē are one—or, to put it another way, that knowing is not set apart from being, but one with being.

This position is opposed to dualism, especially Cartesian dualism, which holds that being is twofold—in other words, that there are two different types of substances, bodies and minds, and thus two different types of things, bodily or material things and mental or thinking things. Cartesian dualism arises when humans come to see themselves as subjects, the specific essence of which is to think, the result being that the idea of thinking or logos in nonhuman nature is rejected, and nonhuman selves are reduced to the same ontological status as entirely mindless material entities. This reduction in turn paves the way for the idea of "man as mentor." Since mind is thought to belong only to humans, it is in humans alone that knowledge and spiritual guidance are to be sought. To affirm the principle of nature as mentor, by contrast, is to affirm that the mental is also the bodily, that there is *no division* between the two, that being and thinking, physis and logos, are one. And since it is thinking (legein) that produces knowledge (epistēmē), it follows that knowledge is also one with being (physis).[34]

The second positive way of characterizing natural knowledge is to say that it is autistic. The key term Maturana and Varela use to describe the cognition of the living is "self-referential."[35] The cognition characteristic of living beings, they claim, does not refer to other entities, but to themselves, by which they mean that it is ultimately directed toward the maintenance of autopoiesis, and not toward, for example, accurate representation of the organism's external environment (as an end in itself). But this is not to say either that biological cognition produces knowledge of the self, for self-knowledge involves treating the self as an object of knowledge. Whatever knowledge is embedded in nature *refers* to the self, but is not *of* the self in the sense of being *about* the self. It follows that the knowledge embedded in living beings has no object; it is not *about* anything, whether another entity or the self. A plant could not know that the chemical composition of water is $H_2O$, for that is knowledge *about* something. Likewise, a plant could not know that it is capable of splitting water into hydrogen and oxygen by means of photosynthesis; such self-knowledge involves treating itself as object. But the process of splitting water into hydrogen and oxygen does involve knowledge of an autistic variety, for it refers back to the self in the sense of being directed toward the maintenance of autopoiesis.

But if knowledge in nature has no object, does it have a subject? If, as I would claim, it is of the essence of the subject to be an "I" that thinks, and if, in thinking, the subject may produce and behold knowledge, then living beings are not subjects, or at least not by virtue of simply being alive. The knowledge embedded in nature is embedded in the process of autopoiesis and, as such, refers back to a self, but this self is not—or, at least, not necessarily—a subject. A self only becomes a subject when referring to itself, when thinking of itself as an "I," or at least when it is capable of doing so. Kant's famous claim that "it must be possible for the I think to accompany all my representations" may hold for all subjects, but it does not hold for all selves.[36]

So, just as natural knowledge has no object, it has no subject either. It is embedded in the way of being of selves, their autopoiesis, but it is not thought by these selves, for if that were the case the self would be more than just a self; it would be a subject. To say, then, that knowledge is embedded in nature is not to say that it is possessed by nonhuman subjects, as if subjectivity were an intrinsic characteristic of nature (at least when legein is present). Knowledge in nature belongs to selves, but not to subjects. What is meant, then, by the claim that the knowledge embedded in nature is autistic is simply that it refers to and belongs to a self. Further, because this knowledge belongs to a self, and not to a subject, it cannot possibly be divided between knowing subject and known object, as is the case when human subjects take nature as object of knowledge. And it is in this sense of belonging to a self, prior to any division between subject and object, that *knowing in nature is one.*

The third way of positively characterizing natural knowledge is to say that it is holistic. We have already seen that knowing is one, in the sense of being, first, undivided from being, and second, undivided between subject and object. But there is another sense in which knowing is one—namely, in the sense of being whole and not divided up into parts, of being unified, rather than individualized into different items. But what exactly does this mean? Natural knowledge acquires significance only in relation to the whole (i.e., the self), to which it both refers and belongs. It follows that it has no independent existence and meaning. This is not to say that natural knowledge is entirely homogeneous. But, since the gathering processes of autopoietic beings belong together in a unified whole, it cannot be said that any individual items of knowledge to which they give rise "exist" within that whole (though they may "ensist" within it). To say that gathering processes refer to the self is to say that they refer to each other in a self-constructing and self-delimiting whole; it is only by means of science that separate items of knowledge may be brought into "existence," in the sense of being artificially delimited from the whole in which they would otherwise only ensist. Prior to the cutting action of science, natural knowledge remains immanent to and concealed within another reality, the natural whole from which it has yet to be delimited.

It could perhaps be objected at this point that the above discussion has drawn extensively on Maturana and Varela, and that their conception of autopoiesis applies only to living beings, or at least only to beings capable of cognition, where cognition is understood in terms of *detecting and responding* to perturbations. This in turn may raise doubts about whether natural knowledge can exist in beings that do not cognize, whether biological (species) or ecological (Gaia). In what follows, I will argue that natural knowledge may indeed arise in the absence of cognition, and that here, too, it may be characterized as autistic, monistic, and holistic.

Let us start by considering species. Now, while species do not cognize, in the sense of detecting and responding to perturbations, they do respond to perturbations. And it is precisely in responding to perturbations, which, as we have seen, takes place by natural selection, that knowledge accumulates within species. Further, a great deal of the knowledge embedded in living beings—knowledge of how to photosynthesize, for example—is more properly considered to belong to the species in that it arises through natural selection and is present throughout the species, which is why it may be abstracted from any member of the species. This knowledge does, of course, also belong to and refer back to biological individuals (and in some cases also to societies), and it is by the scientific study of individuals (or individual societies) that humans may learn from it, but it is not produced by individual (or societal) cognition. Indeed, only knowledge acquired through learning, understood as a cognitive process involving memory, may be said to arise in and be possessed solely by individuals (or societies), and so not also by species. There can be little doubt, then, that knowledge is present in species, and that it is natural selection operating on heritable variations that produces it.

The natural knowledge possessed by species is also clearly monistic, in that it does not belong to some sort of separate mental substance or thing, but is embedded in the "body" of the species itself, understood as an autopoietic entity that here takes the form of a closed reproductive network of biological organisms. It is autistic, for though it refers back to the being of the species, and, as such, is an integral part of its autopoiesis, it is both objectless, in the sense of not being about anything, and subjectless, in the sense that the species is not an "I" that thinks. Lastly, it is also holistic, in the sense that it only acquires the meaning that it has, and so only is the knowledge that it is, in the context of the species in question; it may come to exist as an independent item of knowledge thanks to science, but in so doing it acquires a different meaning, including the meaning it acquires for humans in the context of biomimetic innovation.

As for Gaia, she may also be said to generate knowledge—of how to provide and maintain a habitable earth for her living inhabitants—through processes of selective bringing together or legein. And this knowledge is monistic, in the sense of being both mental and bodily, for there is clearly no reason to think that Gaia has a mind that is in any way distinct from her body. It is autistic, in the sense that it refers back to and belongs to Gaia, understood as that entity the specific characteristic of which is to provide and maintain a habitable earth for her living inhabitants, though without either being about Gaia, understood as an object, or being thought by Gaia, understood as a subject. And it is holistic, in the sense that any item of knowledge we might abstract from Gaia (e.g., knowledge of how to cycle nutrients, accumulate soils, regulate the climate) will assume a rather different meaning once delimited by science, including especially its meaning for us as model or measure.

## Biomimetic Knowledge

We have just seen that the question of the ontological status of natural knowledge may be answered by saying that it is "one," in the sense of being: 1) monistic, and so undivided from being; 2) autistic, and so undivided between subject and object; and 3) holistic, and so undivided into different independently existing items. But biomimetic epistemology is not concerned solely with the knowledge embedded in nature. That knowledge is embedded in nature is a necessary condition for biomimetic knowledge, but it is insufficient. A further condition is that we learn from the knowledge embedded in nature, that this natural knowledge be in some way or other appropriated by humans. Indeed, it is only when we come to consider this further condition that we fully attend to the basic principle of biomimetic epistemology: nature as mentor.

So let us again consider what it means to take nature as mentor. The word "mentor" shares the same PIE root as the words "mind" and "mental"—namely *men-*, meaning "think." But to be a mentor is more than simply to possess a mind or to think. So what exactly is a mentor? The word "mentor" comes from the character Mentor in Homer's *Odyssey*, a friend of Odysseus and advisor to his son, Telemachus. This is why our contemporary word "mentor" has the sense of advisor or spiritual guide. But, in the *Odyssey*, Mentor was often not himself, an ordinary human being, but Athena, the goddess of wisdom, in disguise. Something similar may be said to hold in the case of the principle of "nature as mentor" characteristic of biomimetic epistemology. In biomimetic epistemology, nature is not what it ordinarily appears to be—that is to say, a collection of quite ordinary beings (e.g., plants, animals, forests, mountains) encountered and observed in the world around us, but something quite different and extraordinary: knowledge and wisdom (in disguise).

Further, in biomimetic epistemology, nature also resembles the character of Mentor—when Athena in disguise—in that it is *not itself*, but something *different from and other than itself*—that is, a source of knowledge and wisdom *for others*. It may be of the essence of nature, at least in its biological and ecological instances, to possess thought or mind (logos) and knowledge (epistēmē), but it is not of the essence of nature to be a source of knowledge and wisdom for others. How, then, are we to understand how nature goes from being itself to being a source of knowledge and wisdom for others? And how does the knowledge embedded in nature cease to belong and refer solely to the being in which it was first generated, such that it comes also to belong to other beings—specifically, humans?

With a view to understanding how it is that nature may go from being itself to being a mentor to us, let us consider an analogy. Imagine that there exists a footballer called Plato. Imagine also that it was by watching the legendary Brazilian footballer, Sócrates, that Plato learned how to play. Here we have a situation in which one being teaches another how to do something without doing so intentionally, but simply by going about its own business. Unlike in the case of the ancient Greek philosophers of the same names, Sócrates may not have provided Plato with any pedagogical instruction, never even have sought to offer Plato or anyone else an example to follow, and yet, simply by going about his own business, he was a teacher to Plato, for he showed Plato how to play football. It is in an analogous sense that nature may be a teacher or mentor to human beings. Nature does not provide us with deliberate instruction, does not speak to us, and may, in the case of some species from which contemporary scientists and engineers have learned, not even exist anymore, and yet nature shows and teaches us things in the course of simply going about its own business.

The overall situation here may be expressed as follows. Beings that are in the way of physis produce themselves and, in so doing, show themselves. But while showing themselves belongs to their way of being, showing things to others does not. It follows that physei-beings themselves are not teachers (from the PIE *deik-*, meaning "to show"), but this is not to say that they may not be teachers *to us*. Just as the Brazilian Sócrates was not himself a teacher, but a footballer, and yet, in our imagined scenario, was a teacher to Plato, so physei-beings themselves are not teachers either, and yet they may be teachers to us. Being a teacher or a mentor is made possible by the way of being of physei-beings, for if they did not show themselves in the first place, there is no way that they could show anything to us. Their showing things to us lies not, however, in their way of being, for, as self-referential beings, they themselves could never show things to others, but rather from the way we humans relate to them.

The fact that humans may relate to nature as a source of knowledge implies that our way of knowing is very different from nature's. Indeed, as we will see in what follows, specifically human knowing differs point for point from that of nature: where nature's way of knowing is monistic, autistic, and holistic, ours is dualistic, allistic, and meristic.

Let us start by considering the claim that human knowing is dualistic, by which I mean that with it there emerges a basic ontological duality. Now, this duality is not a Cartesian one; it is not a substance duality between body and mind, between bodily things and thinking things, but between one way of being and thinking, that of nature, and another way of being and thinking, that of humans. The specificity of human being and thinking, as Heidegger recognized, is that it involves understanding things *as things*—for example, understanding water *as* $H_2O$.[37] The duality in question is thus *ontological*, for it institutes a difference between *two different ways in which being is*: on the one hand, being as physis, and, on the other hand, being as the "as," being as that which humans understand beings to be.

And yet the duality is also *mental*, in the sense that it also involves two different ways of thinking (legein): the thinking characteristic of autopoietic beings, which refers only to the self, and the thinking characteristic of human beings, which refers to others (or to the self as object of knowledge) and that, in so doing, understands them "as" something or other, as occurs when I understand myself as, say, a philosopher, or the thing sitting on my desk next to me as a cat. So, whereas Cartesian ontology sees a duality between thinking and bodily things, between mind and matter, enlightened naturalism sees a duality between one way of being and thinking, that of nature, which refers only to the self, and another way of being and thinking, that of human beings, which refers to others (the self included, but as object) and, in so doing, understands them "as" something or other.

Human knowledge also differs from the knowledge embedded in nature in that it is allistic. To say that specifically human knowledge is allistic is to say that it refers to something other (allo-) than the self (autos), as occurs whenever we understand anything "as" such and such. This allows humans to refer to natural beings and thus both to generate knowledge *about* these beings and to acquire knowledge *from* them. These natural beings may include ourselves, but to the extent that we see ourselves as natural beings, we also see ourselves as "other" than ourselves. To make a humanoid robot, for example, we must see the human being as something natural and then seek to learn from that natural entity in various different ways, such as from the form of its hands or from the generative processes its hands accomplish. Human knowledge, it follows, differs from natural knowledge in that it is *divided* between subject and object; human subjects may learn both about and from natural objects.

The third way in which human knowledge differs from that of nature is that it is *meristic* (from the Greek *meros*, meaning "part"). While the knowledge embedded in nature is holistic, and, as such, is held together in an integrated whole, human knowing breaks knowledge up into different items. The meristic character of human knowledge derives from the nature of the human understanding. Whereas in the case of nature, vast numbers of mental or gathering processes will at any point be "ensisting" simultaneously and in parallel throughout the being in question, in the case of the human subject the understanding of things unfolds sequentially and in series, such that one thing is understood after another; water is understood as $H_2O$, $H_2O$ as being split in the process of photosynthesis, the hydrogen thus produced as being a source of energy when recombined with oxygen.

This *meristic* character of human knowledge is also essential to biomimicry. Since human knowing is meristic, it may be applied to natural beings in such a way that, even though their knowing constitutes an integrated whole, to us it may come to appear to be divided up into different items. Trees, for example, may be seen as containing different items of knowledge, such as how to generate usable energy from the sun, how to transport fluids upwards without recourse to a mechanical pump, and how to exchange energy for nutrients with mycelium.

### Biomimetic Technical Knowledge

In the previous sections, I put forward an ontological characterization of both natural and human knowledge, which in turn made it possible to understand how natural knowledge could be appropriated by humans, thus becoming biomimetic knowledge. In this and the following section, I turn to a consideration of the different types of knowledge embedded in nature, as well as, by extension, to the different types of biomimetic knowledge.

We have already seen that natural knowledge is not *about* objects, for it does not consist of true and justified representations, held by a subject, about some object or other (e.g., the representation of water as $H_2O$). So what type of knowledge is it? Two other important types of knowledge not about natural objects are technical knowledge, which is the subject of epistemology of knowing how and epistemology of technology, and ethical knowledge, which is the subject of moral epistemology.

One may perhaps wonder why these were not considered in the above discussion of alternative epistemologies. The reason is simple: technical and ethical epistemologies are not *alternative approaches* to epistemology, but rather *additional branches* of epistemology, concerned with specific types of knowledge: the technical and the ethical. As such, they may be studied from the perspective of the alternative approaches presented above, thus giving rise to, say, naturalized epistemology of technology or feminist moral epistemology.

Recognizing the existence of these alternative branches of epistemology nevertheless points to a specificity of biomimetic epistemology. Whereas the alternative epistemologies discussed above are for the most part concerned with knowledge about natural objects, biomimetic epistemology is not concerned with this type of knowledge, but, rather, with precisely those other types of knowledge that conventional epistemology—its contemporary alternatives included—has a tendency to overlook or to marginalize: technical and ethical knowledge.[38] And in that case, biomimetic epistemology may be understood not only as another branch of biomimicry, understood as a philosophy, but also as that branch that studies the epistemological dimensions of its other two branches, biomimetic technics and biomimetic ethics, including the epistemological dimensions of their relation both to each other and to their common ontological ground in the thinking of being itself as physis.[39]

Let us start by examining technical knowledge. Gilbert Ryle introduced a now famous distinction between "knowing that" and "knowing how."[40] Knowing that is conceived by Ryle as propositional knowledge about some object or other—for example, knowing that water is $H_2O$. Knowing how, by contrast, is characterized as technical knowledge of how to do something or other, such as (to return to a previous example) knowing how to play football. The relevance of this distinction to biomimicry is not difficult to appreciate; while natural knowledge does not take the form of knowing that, it may perhaps be thought to take the form of knowing how, in which case it would be precisely this natural knowing how that is abstracted and transferred over into biomimetic technologies. After all, it seems perfectly reasonable to say that in biomimicry one may learn from spiders *how* to make high-strength fibers, from plants *how* to produce clean energy from the sun, from forests *how* to recycle nutrients, and so on.

Can we say, then, that the first type of biomimetic knowledge that we may glean from nature—the technical type—may be straightforwardly characterized as knowing how? With a view to answering this question, let us first note that the literature on knowing how has had little to say about the question of where knowing how comes from. So, although in the case of traditional epistemology there has been extensive discussion about the sources of knowledge (e.g., the senses, reason, introspection), philosophical discussions of knowing how have almost always taken their cue from Ryle's seminal discussion and have thus concentrated on critical analysis of the distinction between knowing how and knowing that, as well as on the question of whether one of these forms of knowledge is reducible to the other.[41] Moreover, in the absence of systematic philosophical reflection on the sources of knowing how, it seems fair to say that it has been implicitly assumed that knowing how originates in human subjects, whether directly through natural ability, practice, or some combination of the two, or indirectly through learning from other humans.

This is borne out in Ryle's examples of knowing how (e.g., dancing, riding a bike, playing chess, doing mathematics, making good jokes, cooking omelettes, persuading juries), all of which, it would seem, originate either directly or indirectly in human subjects and, as such, do not arise from studying nature.[42] The implicit assumption, in other words, seems to be that one learns how to do things either by oneself or from someone else. That one might learn how to do things from nature is not a possibility that Ryle would appear to have considered.

This is not to say that the basic difference between learning from nature how to do things and learning from other humans how to do things is that nature is composed of "nonhumans." After all, in biomimicry we may also learn from humans, but the way we do so is very different from when we acquire knowing how through imitating their behavior or through receiving explicit instruction from them. There is clearly an important difference between learning from others how to play football, as occurs through observing and imitating other footballers (as in the above example of Plato and Sócrates) or through explicit instruction from coaches, and learning from the scientific study of footballers how to design, say, football-playing robots (or, perhaps more realistically, animated football players in video games).

Analysis of these two different contexts of learning from—learning from other humans and learning from nature—reveals three major differences between them.

The first difference concerns the fact that in nature there is no subjectivity. Ryle, as we saw above, sets out a distinction between knowing that and knowing how. For reasons that I will present later on, however, the true opposite of knowing *how* is not knowing *that* but knowing *about*. And knowing about, in contrast to knowing how, has an object; it does not take the form "S knows that P," where P designates a proposition, but "S knows about O," where O designates an object. The basic grammatical form of sentences expressing knowing about is thus the classic one of *subject verb object*. In the case of knowing how, by contrast, the form is "S knows how to V," where S designates a subject, and V is another verb (e.g., Simon knows how to dance), or, in its underlying grammar, *subject verb verb*. Knowing how thus differs from knowing about in that it has no object. It follows that when we learn from other humans how to do something, such as how to play football or how to dance, the knowledge in question has no object, for it concerns rather something that we do. In this respect, then, learning from other humans is similar to learning from nature.

And yet learning from nature differs from learning from other humans in that in nature there is no subject either. We may sometimes say things like "Plants know how to photosynthesize" (*subject verb verb*), but, since plants are not subjects, this is an incorrect way of speaking. The correct way of speaking about natural knowledge is to say that knowledge of how to photosynthesize is embedded—or, better, that it ensists—in plants. Likewise, when we say things like "Nature knows how to photosynthesize," what this should be taken to mean is that knowledge of how to photosynthesize ensists in nature, not that nature, understood as some sort of metasubject permeating individual beings, knows how to photosynthesize.

The second difference also concerns subjectivity—in this instance, the subjectivity of the learner or mentee. When we learn from another human how to play football, *we* imitate what *they* are doing; one subject imitates another. In the case of biomimetic technologies, however, not only is it not the case that a subject is being imitated but it is also not the case that it is *we ourselves* who are acquiring the knowing how in question. We cannot, for example, learn from plants how to photosynthesize in the way that we may learn from other humans how to play football, for photosynthesizing is simply not something that we humans can learn to do. Rather, human subjects may learn how photosynthesis works in *subjectless nature* and then transfer that knowledge into *subjectless technologies*. Biomimetic technical knowledge, like natural knowledge, is subjectless, even if its production—through the abstraction and transfer of natural knowledge—does of course take place through the faculty of the understanding present in human subjects.[43]

The *subjectless* nature of both natural technical knowledge and biomimetic technical knowledge brings us to a third difference between learning from nature and learning from other humans (as subjects): the form the knowing how abstracted from nature takes when understood by human subjects. To appreciate this difference, we must first revisit a long-standing debate in the epistemology of technology: whether technological knowledge is knowing *how* or just a specific form of knowing *that*. More specifically, the question is whether technological knowledge involves knowing how to do something or knowing that a certain course of action is effective. For example, does the technological knowledge involved in putting a man on the moon involve knowing how to put a man on the moon, or does it involve knowing that a certain course of action (e.g., building a certain type of rocket, launching it from a certain place, propelling it toward the moon at a certain angle and velocity) will enable us to put a man on the moon?

It is not hard to see that in this debate there once again emerges Ryle's distinction between knowing how and knowing that. But, as noted above, the problem with this distinction is that the true opposite of knowing how is not knowing that, but knowing *about*. Once this is recognized, the misleading idea that knowledge must fall into one of two categories—knowing how or knowing that—may be replaced by the following view. There are two types of knowledge at issue here, knowledge of how to do things, and knowledge about objects, and *these may both be expressed in the form of knowing that*: as propositional knowledge. Knowledge *about* the chemical composition of water may be expressed in propositional form—that is to say, in the form "S knows that P" (e.g., I know *that* the chemical composition of water is $H_2O$); and knowledge of *how* to put a man on the moon or *how* to achieve artificial photosynthesis may also be expressed in propositional form (e.g., I know *that* to put a man on the moon or to achieve artificial photosynthesis one must do X, Y, and Z).

From this perspective, the standard opposition between knowing how and knowing that not only obscures knowing about but also fails to recognize that knowing about is something quite distinct from knowing that, just as knowing how and knowing that are also distinct. The result is that knowing how comes to appear as radically different from knowing that, when it is in fact a type of knowing— like knowing about—that may be expressed as knowing that (i.e., in the form "S knows that P").[44] The answer, then, to the conundrum presented to us by epistemology of technology is that *technical* knowledge is knowing how, and *technological* knowledge the expression of knowing how as knowing that—to which one may add that, as such, technological knowledge differs fundamentally from conventional scientific knowledge, which involves rather the expression of knowing about as knowing that.

For present purposes, the importance of resolving this conundrum is that it allows us to see that when the knowing how embedded in nature is understood by humans it is expressed as knowing that—or, to put it another way, that technical knowledge becomes technological knowledge. Knowing how to do something is of course very different from knowing that it is done in a certain way; someone who knows how to play a piece of music on the piano may be unable to explain how it is that they do it, for they may be unable to describe the sequence of notes played. In a similar vein, the first challenge we face when abstracting the knowing how embedded in nature and transferring it over to technology is how to express it as knowing that (i.e., as propositional knowledge). The biomimetic knowledge involved in artificial photosynthesis will derive, first, from understanding the knowledge embedded in nature of *how* to photosynthesize as knowledge *that* usable energy is produced from sunlight by splitting water into hydrogen and oxygen (and other related items of knowledge), and, second, from the development of other forms of technological knowledge to be embedded in biomimetic technologies in the form of subjectless knowing how. For example, knowledge *that* artificial photosynthesis may be achieved by making a certain series of modifications to natural photosynthesis may then come to be embedded in a biomimetic technology, and, through the process of embedding, converted back, as it were, into subjectless knowing how.

In view of all this, it would seem that biomimetic technical knowledge is indeed a form of knowing how. But this conclusion must be substantially qualified, for biomimetic technical knowledge differs from the technical knowledge we acquire from other human subjects when imitating their behavior in three respects: first, we learn it from subjectless nature; second, when we abstract and transfer it we understand it not in the form of knowing how, but rather in the form of knowing that, as propositional knowledge (and, more specifically, as technological knowledge); and, third, once abstracted and transferred from nature it comes to be embedded in subjectless technologies.

*Biomimetic Ethical Knowledge*

In the previous chapter we discussed at length the concept of biomimetic ethics, and thus also the idea that an ethics may be derived from nature. But does it follow that there is ethical knowledge embedded in nature, and that, like technical knowledge, this knowledge may be abstracted and transferred over to humans, thus becoming biomimetic ethical knowledge? Or are there significant differences between these two types of knowledge, in which case biomimetic ethical knowledge would need to be theorized quite differently from biomimetic technical knowledge?

If we are to obtain ethical knowledge from nature, then, because this knowledge derives ultimately from the oikos, we must seek it not in biological entities, such as living beings, societies of living beings, or species of living beings, but in the one true ecological entity, Gaia. This raises the question of the knowledge possessed by Gaia. We saw in the previous chapter that the essence of Gaia, her specific way of being, is to provide and maintain a home for her living inhabitants. It would seem to follow that embedded in Gaia there is *knowledge of how* to provide and maintain a home for these inhabitants—that is to say, knowledge of a technical variety, and that it is this technical knowledge (e.g., how to cycle nutrients, how to sustain life using only solar energy, how to regulate the climate) that may be abstracted and transferred over to artificial entities and systems.

But how, then, does this technical knowledge relate to any ethical knowledge that may be embedded in Gaia? In the case of many autopoietic beings, Gaia included, it is important not to think about the ultimate "end" to which technical knowledge refers back in a linear way. Usually when we think of ends, we think of something at the end of some sort of linear chain, which, as the end of the chain, refers to nothing else. In the case of the autopoietic beings in question, however, any linear chain of references that we may identify within that being is but an artificially delimited part of a circularly organized network in which everything refers either directly or indirectly to everything else. It follows that within these beings there is in fact no end to these chains, for in following them one inevitably goes round in circles. To say, then, that the technical knowledge constitutive of these circular networks refers to the self—that it is self-referential—is not to say that there is an independent entity, the self, to which all the technical knowledge refers back in a linear manner, but rather that it *constitutes* the self, that the totality of these references is what the self is. So, to the extent that it still makes sense to say an ultimate end is present here, it is not some independent thing to which multiple linear chains converge back, but the whole that all the various different items of technical knowledge constitute.

In the case of Gaia, all the various different items of technical knowledge we may abstract and transfer over to human-made technologies ultimately refer back to Gaia's specific way of being, which is to provide and maintain a home for her living inhabitants. And if we inspect Gaia's nutrient cycles, energy flows, and the like, we will indeed find that they together provide and maintain a home for her living inhabitants, for were they either different or inexistent they would not provide and maintain a home for these inhabitants, but either for quite different inhabitants or for no inhabitants at all. But it does not follow that, embedded within Gaia, there lies not just technical knowledge of *how* to do all sorts of different things that together provide and maintain a home for her living inhabitants, but also ethical knowledge *that* providing a home for living beings is the end of all these different items of technical knowledge. Only subjects may know that something or other is the case, and Gaia is not a subject. Similarly, it would be untrue to say that Gaia possesses knowledge about the good or about the ought, as if we could obtain from Gaia some sort of knowledge about ethical matters and then express that knowledge propositionally as knowing that. Gaia possesses no knowledge about either the good or the ought.

Does it follow, then, that there is in fact no ethical knowledge embedded in Gaia after all, that we cannot obtain ethical knowledge from her? Gaia may not know that providing a home for living beings is good, for she knows nothing about anything whatsoever, the good included. But the reason for this is simply that the knowledge ensisting in Gaia, like all natural knowledge, is both subjectless and objectless. Gaia is not a subject that knows something about an object—in this instance, the good or the ought. But this absence of division between knowing subject (Gaia) and known object (the good) at the same time provides a clue to understanding the nature of the knowledge embedded in Gaia. It is not ethical knowledge, in the sense of knowledge *about* the good; rather, it is ecological knowledge, knowledge of how to provide and maintain a habitable earth for its living inhabitants. The key difference between this ecological knowledge and ethical knowledge is that, whereas the latter is *about* the good, Gaia's knowledge *is itself* the good. There is, in short, no division between a subject of knowledge, Gaia, and an object of knowledge, the good; Gaian knowledge is itself the good. And, in that case, to say that in biomimetic ethics we obtain ethical knowledge from Gaia does not mean that we obtain knowledge *about the good* from Gaia, but that the knowledge we obtain from Gaia *is the good*.

So how exactly are we to understand the biomimetic ethical knowledge that we obtain from Gaia? When we come to see that Gaia's way of being and knowing is the good, we come to possess knowledge about the good; Gaia's *ecological* knowledge becomes human *ethical* knowledge. But one may possess knowledge about the good and yet not do the good, whether because of weakness of the will or some other alternative interest, habit, or inclination. What matters, then, is not only to know what the good is, which is Gaia's way of being and knowing, but also to be and to know in the same way as Gaia. If biomimetic ethics involves "being like Gaia," then biomimetic ethical knowledge consists not simply in knowing *that* Gaia's way of being is the good, but rather of "knowing like Gaia"—that is to say, adopting the same way of knowing as Gaia: knowing that enables us to participate in the provision and maintenance of a home for the earth's living inhabitants.

This is not to say, however, that Gaia's way of knowing can be abstracted and transferred over in toto. Technical knowledge may be abstracted from Gaia by human subjects and transferred over into technological products and systems. One cannot, however, abstract and transfer over the whole that is constituted by this technical knowledge. The whole can be understood as such, but not abstracted and transferred over as a whole, for transfer is made possible only by the prior delimitation of parts and the abstraction of models therefrom. The process of delimitation, abstraction, and transfer should, however, be guided by the understanding one has of the whole. To abstract and transfer the natural process of nutrient cycling as a model for the circular economy, for example, should be guided by one's understanding of the ecological whole—in other words, by the way of being of Gaia, which is to provide and maintain a home for her living inhabitants. Indeed, this is precisely what is meant by taking nature as measure: to take nature as measure is to act in accordance with the principle that the "rightness" of what one does, including one's biomimetic technological innovation, is evaluated against the standard set by Gaia.

It follows that it is only models that may be abstracted and transferred, not the basic measure by which our biomimetic innovations are evaluated. To abstract and transfer is in one way or another also to change (even if only in context), and a measure is that with respect to which change may be evaluated, not something that may itself undergo change. If nature's measure were changed by us, nature would no longer be the measure. We may of course fail to live up to the measure or standard set by nature; we may not even *try* to live up to that standard, and in that sense not take nature as measure at all. But, as long as we do take nature as measure, that measure cannot be changed by us, for it is, by definition, set by nature.[45]

To summarize, biomimetic ethical knowledge differs from biomimetic technical knowledge in that it involves knowing in the same way as the Gaian whole, whereas technical knowledge involves abstracting and transferring over to technologies the knowing how embedded in the parts of autopoietic beings, whether biological or ecological. Further, this knowing in the same way as the Gaian whole cannot be abstracted and transferred over to technologies in the manner of technical knowledge, for, in accordance with the principle of nature as measure, its role is rather to guide the abstraction and transfer of technical knowledge as that end to which biomimetic technical knowledge must ultimately refer back.

## BIOMIMETIC SCIENCE

### Conventional Science

In the remainder of this chapter, I will turn from the theorization of biomimetic epistemology to the theorization of biomimetic science. To do this, it will first be necessary to set out a theory of conventional science; just as biomimetic epistemology was initially set out in contrast to conventional epistemology, biomimetic science is best understood by way of contrast with conventional science.

Conventional epistemology sees the epistemic relation between humans and nature in terms of human subjects producing and acquiring knowledge about nature. This conventional epistemology also underlies conventional science, which likewise views nature as the object of knowledge. And yet scientific knowledge is different from non- or prescientific knowledge. Prehistoric humans, for example, undoubtedly possessed a wealth of knowledge about nature, and yet we would hesitate to call it scientific. This raises the question of how conventional scientific knowledge differs from non- or prescientific knowledge, and thus also of what conventional science is.

As we have already seen, the word "science" derives from the PIE root *skei*, meaning to "cut" or to "split." This allows us to see that science involves the attempt to cut nature up into its constituent elements. This first took place in ancient Greece, principally in the work of the atomists, who saw the world as composed of the aptly named "Atoms." These Atoms were believed to have different shapes and sizes and to travel through the void until they collided with other Atoms with which they would sometimes become entangled, thus giving rise to the various composite entities we see in the visible world. In this manner, a model of nature was constructed based on the indivisible elements, or Atoms, and the interactions that took place between them.

But the science of the ancient Greeks differs from that of the moderns. Ever since Edmund Gettier, it has become apparent that the traditional view of knowledge as "justified true belief," which goes back at least to Plato, requires qualification, for not just any old justification will do.[46] Reliabilism, for example, contends that knowledge must be justified using a reliable process. But it was precisely a reliable process of knowledge generation and justification that was lacking in ancient Greece, and it is this lack that differentiates their science from that of the moderns. While theories such as atomism were indeed scientific, in that they sought to cut nature up into its constituent elements, they were also speculative and unreliable; there was simply no reliable way of knowing whether the claims they made were true or not, whether they accurately represented nature or whether they didn't. The justifications offered for one model of nature may have seemed more convincing than those offered for another, but ultimately there was no reliable way of confirming or disconfirming them and thus of establishing them as knowledge. The ancient Greeks may have practiced science, but they lacked scientific knowledge.

The basic reason that theories such as atomism could never be the subject of reliable justification was that the models they put forward were not formulated in such a way that their truth or falsity could be evaluated against measurements made of nature. There was, of course, in atomism a reduction of the qualitative to the quantitative, with qualitative phenomena (e.g., taste, smell, and color) being explained by quantitative differences (e.g., in the shapes and sizes of the Atoms). But the atomists never thought to express the quantitative properties of the Atoms mathematically—that is, with precise numerical values.[47] And, even if they had, the fact that they took over the basic Parmenidean distinction between appearance and reality, according to which reality could only ever be known by the mind, never by the senses, meant that it would in any case have been unthinkable for them to have sought to measure any putative quantities possessed by the Atoms, such as their size or shape. Everything they had to say about these sizes and shapes was thus entirely speculative. When Democritus tells us that jagged Atoms taste bitter whereas round Atoms taste sweet, this is pure speculation.[48]

So, while the ancient Greeks put forward models of reality, these were never able to attain the status of scientific knowledge, for it was impossible either to confirm or disconfirm them on the basis of measurements. This is not to say that the ancient Greeks did not measure things. Measurements were made at the macroscopic level of everyday reality—for example, when weighing goods or measuring the lengths of geometrical figures. But these measurements were not—and, in the case of the atomists, could not—be used to evaluate the truth of their models of nature. In summary, then, we may say that the ancient Greeks possessed models of nature, as well as the ability to make measurements, but they did not articulate the two and were thus unable to produce scientific *knowledge* about nature.

What enabled the emergence of scientific knowledge in modernity was precisely the possibility of putting forward models whose truth or falsity could be evaluated by taking measurements. Underlying this change lay two key developments. First, there was the idea that nature is expressible in the language of mathematics, that it can be precisely and rigorously quantified.[49] Second, there was the idea that the interactions of the basic elements of nature—the Atoms—were governed by forces. So, while the ancient view of Atoms as solid structures capable of becoming entangled with one another meant that no further phenomena were considered necessary to explain their interactions, in modern science these interactions are explained by forces. Further, since these forces can act on aggregates or compositions of Atoms, it is not necessary directly to perceive the Atoms in order for the model to be evaluated by means of measurements; the effects of forces acting on aggregates or compositions of Atoms may be visible even if the Atoms themselves are not. Isaac Newton's theory of gravity could be tested by measuring macroscopic phenomena—the movement of the planets, for example—without directly observing and measuring the individual Atoms.

Taken together, these two developments meant that models of nature in terms of interacting elementary units and expressed in terms of equations could be tested by taking measurements of nature and then putting the results obtained into the equations in such a way that the model could be either confirmed or disconfirmed.[50] Modern science, it follows, operates according to an *iterative interplay of models and measures*. Models are evaluated on the basis of measurements, such that any discrepancy between the model and the measurement (assuming the measurement to be accurate) requires the model to be either revised or abandoned and a new model put forward in its place. This new model may in turn be either confirmed or disconfirmed by taking further measurements, and so on, potentially ad infinitum. So, while conventional science, the fundamental characteristic of which is to produce models of reality that describe the interactions of the fundamental material elements, goes back to the ancient Greeks, it underwent a revolution during the modern period, such that it is now able to produce genuine scientific *knowledge* by evaluating the truth of the models proposed against measurements of nature.

## Biomimetic Science

The preceding discussion of conventional science made no great claim to originality. Its aim was simply to present what I believe to be a relatively uncontroversial understanding of conventional science in a way that allows us to see how biomimetic science is both analogous to and—because of its epistemological orientation toward learning *from*, rather than learning *about*—quite different from conventional science.

The biomimicry of the ancients resembled their science in the sense that it began with cutting procedures. These took two forms. First, beings were cut up into different *parts* from which models could then be abstracted. To take a famous example from ancient mythology, Daedalus imitated the wings of birds in order to facilitate his escape from imprisonment in a tower by King Minos of Crete. Second, different types of *traits* were cut out or abstracted from these beings or systems. The bronze statue described by Aristotle imitated the form, but not the materials, of the human being on which it was based. Or, to take an example from Democritus, human-made hunting nets were thought to have emerged through imitating the form and function of spider's webs, but not their materials.

Just as ancient science differs from modern science in two key respects, so the same is true of ancient biomimicry. First, nature is not mathematized. The forms of human beings and spider's webs may well have been taken as models for human-made artifacts, but mathematization was absent both from the study of nature and from the production of artifacts modeled on nature. Second, the interactions between the beings from which models were abstracted were not considered to be governed by any sorts of laws or principles. Just as there were no "forces" in ancient atomism governing the interactions between the Atoms, only the Atoms themselves interacting with one another on the basis of their own intrinsic properties (their shapes and sizes), so the same was true of ancient thinking about living beings. Living beings went about their own business, interacting with one another on the basis of their intrinsic properties (e.g., carnivores killing herbivores with their sharp teeth and claws), but there were no specifically ecological principles governing their interactions, and thus no attempt made to engage in the study of, for instance, nutrient cycles, energy pathways, and predator-prey population dynamics, or, a fortiori, to quantify these ecological phenomena in the language of mathematics.[51]

These two changes—the mathematization and ecologization of nature—underlie the transition from ancient biomimicry to contemporary biomimetic science; they make it possible to take nature not only as model, as was already the case in ancient biomimicry, but also as measure. This is not to say that the idea of nature as measure was completely absent from ancient biomimicry. Nature was no doubt seen as setting a standard that humans could not surpass—an idea also present in the Middle Ages, when the perfection of God's creation set the unsurpassable standard against which human imitations were measured, and in the modern period, when the unsurpassable standard was set by Man himself. But what mathematization and ecologization made possible was the idea that technologies based on natural models could be *precisely measured against an ecological standard set by nature*. Just as the mathematization of the interactions between the Atoms makes possible modern conventional science, so the mathematization of the interactions between living beings (and their abiotic environments) makes possible contemporary biomimetic science.

Conventional and biomimetic science may thus be expressed in much the same terms—that is to say, in terms of the *iterative interplay of models and measures*. And yet, as will soon become apparent, thanks to a wonderful ambiguity of language what the word "model" signifies in these different contexts is diametrically opposed. In conventional science, "model" usually refers to a representation of some sort of system, understood as a set of interacting elements. In the broader sense in which I propose to use it, however, it may refer either to a model or simulation of a specific natural system (e.g., a model or simulation of the solar system) or to a hypothesis or theory applicable to multiple different natural systems (e.g., the theory of gravitation), the primary difference between the former cases and the latter being the level of abstraction. Newton's theory of gravitation is a highly abstract model in that it may be applied to any system of entities possessing mass and situated relative to one another in space. A model of the solar system, by contrast, is of a lower degree of abstraction inasmuch as it applies to one specific set of entities, the sun and orbiting planets.

In biomimetic science, by contrast, the word "model" refers to some sort of blueprint or template to be imitated by human technology. So, whereas in conventional science the model is the imitation of the thing, in biomimetic science it is rather the thing to be imitated. This is why I earlier spoke of an "ambiguity" (from *ambi*, meaning "both," and *agere*, meaning "lead") of language; the word "model" may lead both toward and away from nature.[52] In conventional science, the model *leads toward nature* as that which it is to represent. In biomimetic science, the model *leads away from nature* as that which is to be imitated by human technology. This may be represented by the following schema, in which the first arrow, turned toward nature, represents the epistemological orientation of conventional science, and the second, turned away from nature toward technology, the epistemological orientation of biomimetic science (see fig. 4.1).

Moreover, just as in conventional science the model may be more or less abstract, with the word "theory" often given to the more abstract of scientific models, so in biomimetic science the model may also be more or less abstract (as we saw in chapter 2). One may, for example, abstract from nature either the "level 1" model of the self-cleaning surface of the lotus plant or more abstract "level 2" models of self-cleaning surfaces in general and that may derive from observing multiple different instances of self-cleaning surfaces in nature.

In keeping with these opposed epistemological orientations, the word "measure" is also used quite differently in conventional and biomimetic science. In conventional science, a measure or standard is some sort of constant, relative to which measurements (i.e., determinations of magnitudes) may be made, and that thereby makes possible the evaluation of representational models of nature. Some of these standards or measures are anthropocentric in that they derive from the everyday world of humans, as is the case when distances are measured in feet and inches, but also, albeit to a lesser extent, when measured in meters, for to see a meter as a basic unit of distance is to choose a unit that is both purely conventional and readily perceived and understood by humans. But there are also

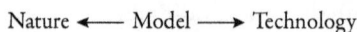

Nature ◄─── Model ───► Technology

4.1 The model as representation of nature and as template for technology nonanthropocentric measures in the form of fundamental physical constants, believed to be both universal in nature and constant in time, such as the speed of light or the gravitational constant. What is common to these measures—whether anthropocentric or nonanthropocentric—is that they make it possible to take measurements and thus to evaluate the truth of the models we make of nature. In the last analysis, however, the ultimate measure against which our scientific models are evaluated is none other than physical reality itself, for it is physical reality that provides the fundamental and ultimate measure on the basis of which all other measures or standards are derived, all measurements made, and all scientific models evaluated.

In the case of biomimetic science, by contrast, the measure is again the standard relative to which measurements are made, but what these measurements ultimately allow one to evaluate is not the rightness of the model (qua representation), but rather the rightness of the technology based on the model (qua template). As in the case of conventional science, it is possible to adopt an anthropocentric measure of a technology's rightness. If the standard concerns the contribution of the biomimetic technology to economic growth or to aggregate human well-being, the measure is anthropocentric. But, as we have already seen, a nonanthropocentric measure may also be identified in nature—namely, the way of being of Gaia, which is to provide and maintain a home for her living inhabitants. A biomimetic technology may thus be judged according to the extent to which it approximates to—or veers away from—the ecological standard set by nature. This nonanthropocentric measure differs from the measures of conventional science in that it is not a *physical* measure taken from the basic properties of physical reality, but rather an *ecological* measure taken from the basic properties of the terrestrial oikos.

If the ultimate standard or measure in biomimetic science is the way of being of Gaia, then it is also true that one may identify various submeasures within Gaia relative to which the ecological performance of our technological systems and products may be evaluated. When these submeasures are used to evaluate technological systems modeled on natural ecosystems, they concern such phenomena as nutrient cycling, renewable energy generation, and air and water purification. When they are used to evaluate technological products modeled on biological systems, they concern such phenomena as recyclability, ability to generate or run on renewable energy, and ability to purify or at least not pollute air and water. To take an example of the former, one may evaluate the ecological performance of a city by comparing it to that of the indigenous natural ecosystem (e.g., with respect to nutrient cycling, renewable energy generation, air and water purification), with a view to transforming the city in such a way that, thanks at least partly to the imitation of natural models abstracted from that ecosystem, the city may more closely approximate to its ecological standard.[53] And, to take an example of the latter, one may evaluate the ecological performance of a self-cleaning surface modeled on the lotus plant by measuring its ecological performance against that of the lotus plant itself (e.g., with respect to recyclability of materials, whether it was made using renewable energy, whether its fabrication or use is polluting).

So, just as in conventional science, there is one fundamental measure, physical reality itself, which may in turn be analyzed in terms of various different submeasures (e.g., units of time, speed, distance, gravitational attraction, electric charge), on the basis of which quantitative measurements may be made and that in turn make possible the evaluation of scientific models, something analogous holds in the case of biomimetic science. The one fundamental measure—providing and maintaining the habitability of the earth for its living inhabitants—may be divided up into various different submeasures (e.g., related to nutrient cycling, renewable energy generation, air and water purification), on the basis of which quantitative measurements may be made and that in turn make possible the evaluation of technological products and systems based on natural models.

In its broad outlines, then, the iterative interplay of models and measures constitutive of biomimetic science proceeds as follows. A model is first abstracted from a natural system and a new technology based on that model developed (or at least conceived). This technology—or perhaps a model or prototype thereof—is then evaluated against the ecological standard of the natural system, bearing in mind that this standard covers not only effectiveness (and efficiency) but also appropriateness and sustainability, such that the question is not only whether the technology works as well as the natural system (e.g., whether an artificial self-cleaning surface is as effective as a natural one at removing particles of dirt), but also whether it is appropriate (e.g., whether it is harmless to local wildlife) and sustainable (e.g., whether it is recyclable). If the technology is found to deviate significantly from the ecological standard set by nature, it may either be upheld as a temporary solution on the grounds that it is better than other available solutions (just as scientific models may be temporarily upheld despite their known deviations from relevant measurements), or it may be immediately revised such that it more closely approximates to the standard set by nature.

To illustrate the iterative interplay of models and measures characteristic of biomimetic science, let us consider the cases of energy generation and storage. There can be little doubt that much of the impetus behind the research, development, and use of photovoltaic solar panels has come from explicit awareness of the fact that, like nature's plants, solar panels produce energy from the sun. The technology that currently dominates this field, photovoltaics, may thus be categorized as biomimetic in the sense that a function abstracted from nature—producing usable energy from the sun—has provided a significant source of inspiration for the technology's development and deployment.

But if one compares the effectiveness, appropriateness, and sustainability of photovoltaics to that of natural photosynthesis, it fails in many respects to live up to nature's ecological standard. Indeed, it would seem that while photovoltaics may compare favorably to natural photosynthesis in terms of effectiveness, for energy conversion rates are typically higher, the comparison is much less favorable when it comes to appropriateness and sustainability.[54] Natural photosynthesis uses locally available biodegradable materials, and in this respect may be considered highly appropriate, in the sense of fitting into the contexts in which it is embedded. Photovoltaic panels, by contrast, require the use of materials that are often not available locally, including various rare earth metals available only in small quantities and in a limited number of global locations.[55] Further, whereas the leaves and other organs in which photosynthesis takes place may be broken down into simple materials at ambient temperature and then reused by plants or other organisms and, as such, are perfectly sustainable, photovoltaic panels raise significant sustainability issues, in particular as regards their environmental impact,[56] and the recoverability of the materials used.[57]

Measuring photovoltaics against the ecological standard set by natural photosynthesis may thus lead us to modify photovoltaic panels in such a way that they approximate more closely to natural standards of effectiveness, appropriateness, and sustainability.[58] It may, for example, be possible to make use of low-energy manufacturing processes, to reduce the need for toxic materials, or to design panels that are more easily recycled.[59] It is also possible, however, that improvements to photovoltaics such as these may reach a point beyond which it proves difficult or impossible to advance. In these instances, biomimetic science calls for a modification of the model on which the biomimetic technology is based. In the case of solar energy generation, this could involve the development of a new technology that imitated not only the *function* of producing usable energy directly from the sun but also the generative processes, forms, and materials involved in natural photosynthesis. This new technology could then be evaluated with respect to the measure provided by natural photosynthesis and then either modified so as to make it approximate more closely to nature's ecological standard, or, alternatively, the abstracted model (or models) it imitates could undergo further modifications with a view to producing another more ecologically benign technology.

Consider also the case of energy storage. Conventional technology characteristically uses energy already stored in some way or other (e.g., uranium, coal, oil), the result being that the question of how to store energy does not emerge as a significant engineering problem. Once we start imitating nature in using natural energy flows, however, we are faced with the problem of how to do something that nature has already learned how to do: store energy. Energy storage technologies, it follows, may be thought of as biomimetic inasmuch as they seek to realize a function already present in nature: the storage of energy derived from natural energy flows, especially sunlight.

At present, the dominant technologies in this field are electrochemical batteries (especially lithium-ion batteries). As in the case of solar panels, however, these compare unfavorably to nature's ecological standard, for they raise problems regarding such things as recyclability, availability of components, and production of harmful wastes.[60] Awareness of these ecological deficiencies may provide the motivation to adopt a model that would allow energy storage technologies to approximate more closely to nature's ecological standard. An obvious candidate is to store energy in the form of hydrogen, as occurs in nature. But this may in turn only allow us to get so close to nature's ecological standard. It may be that problems involved in the storage of hydrogen in a pure form (as a pressurized gas or very cold liquid) create further problems, in which case it may be necessary to revise the model again—for example, by not simply seeking to store hydrogen, but, more specifically, to do so in the form of chemical compounds, as in nature.[61]

The above examples also reveal something important about the contexts in which biomimetic science occurs. Biomimetic science may, of course, occur at the level of a specific project of technological development—for example, a technology being developed in the field of artificial photosynthesis, which, through the interplay of models and measures, may over time come to approximate more closely to the ecological standard of natural photosynthesis. But biomimetic science may also occur at a much broader, societal level, with competition between technologies being driven by the goal of approximating to ecological standards. Some artificial photosynthesis technologies could approximate more closely to nature's ecological standards than others and thus out-compete others. The same principle could also hold between radically different technologies; for instance, wind turbines and artificial photosynthesis technologies may both undergo a series of improvements through the iterative interplay of models and measures, but there may reach a point where one of these technologies clearly emerges as superior to the other, as measured by proximity to nature's ecological standard and thus comes to dominate, at least in the majority of contexts.

Thus, just as in conventional science, different models may develop side by side (e.g., different hypotheses regarding the nature of dark matter) before one or the other emerges triumphant, so the same is true of biomimetic technologies. It is quite possible, for example, that the iterative interplay of models and measures would tend, over time, to favor artificial photosynthesis and hydrogen storage over other forms of renewable energy generation and storage, such as wind turbines and electrochemical batteries. But it is also possible that each of these technologies may develop according to the iterative interplay of models and measures, while also adapting to a specific context, and so without entering into direct competition with the other, just as the theories of conventional science may avoid competition by covering different parts or aspects of physical reality.

An important question raised by the above analysis of biomimetic science is whether it means that biomimetic technologies would, over time, come to resemble ever more closely the natural systems on which they are based. Indeed, it may be thought that this is precisely what is suggested by the analogy with conventional science, for it would no doubt appear obvious to many that, as conventional science progresses (via the iterative interplay of models and measures), the models it produces come to represent physical reality ever more accurately. For a variety of reasons, however, this conclusion does not follow from the analogy with conventional science.

The first reason concerns the fact that the models of conventional science do not always seek to represent ever more accurately. They also seek parsimony, an economical description of the phenomena, and this in turn often leads to abstraction, with certain features of the underlying physical reality being overlooked or simplified in the process of elaborating the scientific model. So, while scientific models are improved by more closely respecting the standard set by nature, in the sense of more closely conforming to the measurements made, the tendencies toward parsimony and abstraction mean that the model will not necessarily seek to represent the totality of the natural system being modeled. A model of solar system dynamics may, for example, be improved by more accurately representing the masses and velocities of the planets, but this is not to say that it should necessarily seek to represent the materials of which the planets are made; if the model serves only to predict the movements of the planets, it may be that only their mass, not their material composition, is relevant.

Something similar applies in the case of biomimetic science. The models we abstract from nature often exhibit parsimony and abstraction. A model may ignore the material of which the natural system is made, retaining only its form, as in the case of the high-speed Japanese train modeled on the form of the king-fisher's beak. Similarly, a model may reproduce only some of the most general characteristics of a natural form, common also to certain other entities, and not the precise details of the natural form specific to the entity from which it was abstracted.

A second reason concerns the debate between positivist and postpositivist philosophy of science. Positivist philosophies of science typically assume that, over time, scientific models (theories included) come to represent nature ever more accurately. Postpositivist philosophies of science have questioned this assumption. Thomas Kuhn, for example, has argued that the history of science is the history of scientific paradigms which succeed each other without necessarily tending toward an ever more accurate representation of nature.[62] Something analogous may perhaps be the case in biomimetic science. It could be that increased awareness of "anomalies" with respect to nature's ecological standard will drive a series of technological revolutions, and thus also a succession of technological paradigms, but that the overall trend of these revolutions will not be toward ever more exact imitation of natural systems.

Consider the case of aviation. Dessauer was not wrong to say that mechanical flight was first achieved when the model of flapping wings was abandoned. This is not to say that the natural model was entirely abandoned, for many of the principles applied in mechanical flight derive from the study of bird flight.[63] But it is nevertheless true that a significant part of what made mechanical flight possible was the rejection of a model previously abstracted from nature (flapping wings) in such a way that, at least in that one respect, the technology came to resemble the natural system *less closely than before*.

Of course, it could perhaps be argued that the significant difference thus introduced from nature was problematic, for jet engines running on fossil fuels are ecologically damaging in a way that flapping wings running on the nutrients consumed by birds are not. A gain in effectiveness may have been made at the cost of a loss of appropriateness and sustainability. One might also add here that, aware of the damaging ecological impact of flying, aviation is once again looking to birds as a source of models.[64] But this is not to say that flapping wings will one day necessarily be reintroduced, or that if other models, such as flying in formation, were to catch on they would all necessarily survive future ecologically driven paradigm shifts in aviation technology.

In the last analysis, what matters in biomimetic science is not exact reproduction or duplication of the natural system, but approximating to the ecological standard, which ultimately refers back to the provision and maintenance of a habitable earth for its living inhabitants. And it may well be that ever closer imitation of natural systems is not necessary to achieve this, that the iterative interplay of models and measures may lead to a succession of biomimetic technologies—and perhaps also a succession of paradigms in biomimetic technology—that in different respects and at different times tend *both toward and away from* the natural systems on which they are based.

The third reason biomimetic technologies may not come to resemble ever more closely the natural systems on which they are based is that there will in numerous instances be limits to which this is even possible given their respective contexts. In a great many cases, biomimetic technologies exist primarily for the benefit of human beings, and, as such, the context in which they find themselves is somewhat different from that of the natural system on which they are based. A passenger plane, for example, is necessarily quite different from any bird because it must be able to transport human beings, and this in turn means that it will necessarily differ from birds in such terms as size and internal arrangements. Likewise, a building may be like a tree in generating energy from the sun, being made primarily out of wood and facilitating rainwater infiltration, but it will also need to have sufficient internal space for its human inhabitants, and that internal space will in turn require a form that differs quite significantly from those of trees. This is not to say that man is the measure of biomimetic technologies after all. If a home is to be provided and maintained for life on earth, then that includes all humans alive today, and it will thus be necessary to provide the artificial habitat that we require in order to flourish—subject to the proviso that this arti-ficial habitat participates in the provision and maintenance of a habitable earth for nonhumans too.

In view of all of this, it seems to me an open question as to whether progress in the field of biomimetic technologies—where progress is understood as the devel-opment of technologies that approximate ever more closely to nature's standards of effectiveness, appropriateness, and sustainability—would generally tend toward ever closer imitation of the natural systems on which biomimetic technologies are ultimately based. To respect nature's ecological standard will no doubt require us to develop biomimetic technologies that *in some respects* more closely resemble the natural systems on which they are based, but this is not to say that these tech-nologies will have to resemble the natural systems more closely *in every respect*. In some instances, it may turn out to be more effective, appropriate, and sustain-able to develop technologies that differ in certain key ways from natural systems.

*Interactions Between Conventional and Biomimetic Science*

In the preceding discussion, I presented and discussed an analogy between con-ventional and biomimetic science. But while two things may be analogous with-out interacting, this would not appear to be the case here, in which case a full understanding of the relationship between conventional and biomimetic science requires us also to consider the nature of their interactions and not just their for-mal similarities and differences.

It may at first sight seem to be the case that, if we are to learn *from* nature, we must first learn *about* nature, that conventional science must precede biomimetic science. From this perspective, we must first model nature, before proceeding to take that model (in the sense of a representation) as a model (in the sense of a template) for a biomimetic technology. We must, for example, first develop a model of natural photosynthesis (in the sense of a representation), which we may then take as a model (in the sense of a template) for artificial photosynthesis. A similar point may appear to apply also to measures. We must, for example, first measure the levels of nutrient cycling in a natural ecosystem just as we would in conventional science, before taking the measurements thus obtained as a stan-dard against which levels of nutrient cycling in the artificial system are evaluated. Conventional science, viewed in this way, has both logical and methodological priority, with biomimetic science assuming a secondary and dependent status.

There is, however, another way of looking at the interactions between conventional and biomimetic science. To see this, consider the contrasting language one uses to talk about models in these two contexts. In conventional science, one talks about "modeling" nature, or "putting forward models" of nature. In biomimetic science, by contrast, one talks of "abstracting models" from nature. This points to the truth that modeling nature and abstracting models from nature are different courses of action: one *either* proposes a model of nature *or* one abstracts a model from nature. In keeping with this, the models produced by conventional and biomimetic science may be quite different; because they have different epistemological orientations and objectives, they may focus on different properties of the natural system in question and may seek parsimony and abstraction in different ways. Something similar is also true regarding measures and measurements. The measures selected and the measurements made to evaluate the rightness of representational models of nature may be quite different from the measures selected and the measurements made in order to evaluate the rightness of biomimetic technologies.

Does this imply, then, that conventional and biomimetic science do not interact after all, that they simply pass each other by as they each go about their own business? Not necessarily. It is certainly possible that a scientific model produced in order to represent nature could be *taken as* a model for a biomimetic technology, whether as it is or modified in some way or other—for example, by removing details that are extraneous in the latter context. But it is also possible that a model abstracted from nature for the purposes of biomimetic design could be used, either as it is or in modified form, for the purposes of conventional science.

The same principle applies to measures and measurements. Measurements made of a natural ecosystem in order to evaluate a representational model in conventional science could be taken as standards against which an artificial system is evaluated. Conversely, measurements made of a natural ecosystem in order to evaluate an artificial system modeled on it could potentially be used to evaluate a representational model in conventional science. We may conclude from this that, while there are no necessary interactions between conventional and biomimetic science, there may be contingent interactions. Models and measurements produced in each of these different scientific contexts may be appropriated by the other, either as they are or modified in some way according to their differing epistemological orientations and objectives.

We may also conclude that there is no necessary priority of either conventional or biomimetic science. We considered earlier the plausible claim that biomimetic science is secondary with respect to and dependent on conventional science. But it would not be out of the question to make precisely the opposite case. Vico's *verum factum* principle—according to which one only fully understands something when one is able to make it oneself—could be invoked here in order to argue that it is only by making something artificially that one comes truly to understand the natural phenomenon on which it is based. Only by achieving photosynthesis artificially, for example, would we truly understand natural photosynthesis itself.[65] But, while it would surely be mistaken to think that it is in the very nature of conventional science to be secondary and dependent on biomimetic science, there may be contingent instances in which the relationship is this way round, when advances in conventional science derive from biomimetic science, whether from the abstraction of a model or its successful transfer into a technology, just as there are clearly cases where biomimetic science is made possible by knowledge about nature generated in the context of conventional science.

Although there is no necessary priority of either conventional or biomimetic science, there may still be a contingent priority relation. Indeed, there can be little doubt that over the past two or three hundred years, conventional science has advanced significantly, whereas biomimetic science has been made at best peripheral, and at worst practically impossible, by the prevalence of alternative or incompatible technical, ethical, and epistemological perspectives, in particular the widespread view, common to philosophy of technology as it first emerged in the nineteenth and early twentieth centuries, that technology *is not* imitation of nature. It is also true that the vast quantities of knowledge about nature accumulated over this period have played—and will continue to play—a major role in the advance of biomimetic science.

Nevertheless, it is also possible that the radical ontological, technological, ethical, and epistemological shifts that together constitute the biomimicry revolution may give rise to a reversal of the current priority relation between conventional and biomimetic science. It is, in short, quite possible that the primary epistemological orientation of humankind will in the future be toward learning from nature, rather than learning about nature, and thus also toward biomimetic science as opposed to conventional science. Were technology seen as imitation of nature, and ethics as emulation of nature, then it is possible that biomimetic science and epistemology would be given priority with respect to conventional science and epistemology, that models and measurements made in the first instance for the purposes of biomimetic science would thereafter be appropriated by conventional science more often than the other way round. As a consequence, natural scientists would more often find themselves teaming up with engineers, designers, architects, and the like, with a view to developing biomimetic products and systems, than they would either working on the stand-alone pursuit of representational truth or on forms of applied science that seek neither imitation nor emulation of nature.

Biomimetic epistemology, in this scenario, would aim at more than just a theory of biomimetic knowledge. It would also aim at a biomimetic theory of knowledge—that is to say, a theory according to which knowledge arises primarily, though certainly not exclusively, through learning from nature.

---

The basic insight of biomimetic epistemology—that we may view nature not only as an *object* of knowledge but also as a *source* of knowledge—may, I believe, come to underlie a new relation with nature.

At present, thinking about the relation between humans and nature increasingly focuses on the postmodern or posthumanist idea that nature possesses agency. In place of the modern view of humans as agents manipulating and controlling passive natural objects, there has thus emerged a view of humans and nonhumans as participants in actor-networks capable of responding to one another in complex and often unpredictable ways.[66] To the extent that enlightened naturalism sees those natural beings in which legein is present as capable of responding to perturbations, it agrees with the characterization of nature as possessing agency. Biomimetic epistemology differs from postmodernism and posthumanism, however, in three important respects. First, its initial focus is not on the agency of nature, but rather on the presence of mind—in the sense of legein—in nature. Second, while it sees agency as a consequence of legein, it is more interested in how legein produces knowledge in nature than with agency per se. Third, this focus on knowledge in nature makes it possible to see the human relation to nature in terms of learning from the knowledge first generated in nature by various different modes of legein (e.g., natural selection, cognition, intelligent cognition), rather than in terms of—or primarily in terms of—the interaction of human and nonhuman actants.

This new epistemological relation to nature is also of major practical significance. Indeed, while it is not always very clear what difference seeing nonhumans as actants might make to how we conceive such practical branches of thought as technics and ethics, the same cannot be said of the new relation to nature that may emerge from seeing it as a source of knowledge.[67] According to Benyus, the biomimicry revolution differs from the industrial revolution in that "this time, we come not to learn *about* nature so that we might circumvent or control her, but to learn *from* nature, so that we might fit in, at last and for good, on the Earth from which we sprang."[68] So, whereas Heidegger tells us that in the era of modern technology nature is seen as "standing-reserve" (*Bestand*)—that is to say, as a set of resources to be "ordered" (*bestellt*) by human beings—in the new era introduced by the biomimicry revolution, nature is no longer seen as simply the passive recipient of human ordering, but rather as a *source of order* that humans must respect. No longer is it a question of rearranging nature however we see fit, but of trying to fit into and become participants in the natural order, the ecological way of being of Gaia.

But this practical objective is not just a vague ambition; by means of biomimetic science, it may be operationalized in a precise and concrete manner. If the basic measure to be respected is always the fundamental ethical one of participating in the provision and maintenance of a habitable earth and thus "fitting in" on earth, this measure may also be divided up into various submeasures, some of which may be specific to sub-Gaian ecosystems. This in turn makes possible the *quantitative evaluation* of the performance of our products and systems against those of nature.

This practical objective also provides a somewhat critical perspective on many contemporary attempts to be "green." As we increasingly turn the earth inside out in search of ever more rare metals to power our solar panels and electric vehicles, we may do well to remember that nature offers us the tantalizing prospective of radically different ways of doing things (e.g., different materials, different manufacturing processes, different structures and organizations). This is not to say that the imperative of avoiding irreversible climate change does not justify using these low-carbon technologies in the short-term, but in the long run there is much scope for potentially very different technologies that would allow us to approximate more closely to nature's ecological standard.

In view of this, it is clear that biomimetic science offers not only a new method for ecological innovation but also a new conception and yardstick of progress.

As such, it allows us to glimpse what a truly sustainable long-term human habitation of the earth might involve. If, as Heidegger rightly says, we must "ever learn to dwell," then it is also true that, thanks to biomimetic epistemology and science, we may potentially find answers to the challenges we face in our attempts to dwell today, as well as an overall vision—the concrete details of which remain undecided—of how our future dwelling might unfold.[69]

# Conclusion

## Toward a New Enlightenment

There is a broad assumption shared by many that biomimicry is simply a strategy for technological innovation, albeit one oriented toward sustainability. Over the past few years, a small number of philosophers have sought to question this narrow and reductive interpretation of biomimicry, suggesting that to embrace biomimicry is to embrace not only a new strategy for technological innovation but also a broader philosophical shift in both our thinking about and our relation to nature.

When I started working in this area, however, this insight had yet to be developed in such a way that biomimicry might come to be seen as a genuinely new philosophy. This is what the present work set out to accomplish: not so much a *philosophy of biomimicry*, understood as an attempt at sustained philosophical reflection on the subject of biomimicry, as the exposition and development of biomimicry, seen as a *new philosophy*. From this perspective, biomimicry as I initially encountered it was not so much a fully fledged entity awaiting philosophical analysis, clarification, and problematization, but more an embryonic collection of principles and insights that had yet to be understood and developed into a coherent philosophical system.

The biomimicry revolution is, however, far from being only a philosophical revolution, for the philosophical foundations of biomimicry—in ontology, technics, ethics, and epistemology—imply radical changes to human being, producing, acting, and knowing. But if the biomimicry revolution is not simply a philosophical revolution, and if its ramifications beyond philosophy encompass more than just a technological revolution, what sort of revolution is it? I will conclude by sketching out the idea that the biomimicry revolution is best seen as—or at least as making an essential contribution to—a new enlightenment.

The question "What is Enlightenment?" was famously asked by Kant. His answer was that "enlightenment is man's emergence from his self-incurred immaturity." And immaturity, he added, "is the inability to use one's own understanding without the guidance of another."[1] At the center, then, of Kant's understanding of enlightenment is the concept of autonomy. To be enlightened is to depend no longer on the "guidance of others," for that would be heteronomy, but rather to think for oneself, to be autonomous. Expanding this focus on autonomy beyond "What is Enlightenment?" to include Kant's philosophical system as a whole, we may say that Kant thought autonomy manifested itself at four different levels: ontology, technics, ethics, and epistemology.

At the ontological level, Kant believed, though did not claim to know, that humans were free, that beyond the realm of empirical reality in which everything is determined by something else lies the realm of transcendental reality, in which the human being is capable of self-determination, and, as such, free. At the level of technics, Kant thought that freedom manifested itself in the form of creativity. To be free is to come up with one's own works of art (and perhaps technology), to invent things through one's own genius. At the ethical level, human freedom manifests itself not in the form of arbitrary choices, but in letting oneself be determined by another essential human trait: pure reason. And, at the epistemological level, freedom manifests itself as embracing the opportunity to use one's in-built faculty of understanding in conjunction with sensory intuition in order to produce knowledge, especially scientific knowledge, about empirical reality. In sum, Kant saw humans as ontologically free, and, as such, capable of freely generating works of art and technology, ethical and political codes and systems, and corpuses of knowledge, especially of the scientific variety.

At first sight, the biomimicry revolution may appear to be in radical contradiction with this Kantian view of enlightenment. After all, if to be enlightened is no longer to need the "guidance of another," then to turn to nature for guidance, as biomimicry does, may appear to be a return to immaturity. But this is not quite right. Immaturity, Kant tells us, is not turning to another for guidance, but rather the *inability* not to do so, the inability to think on one's own. And this ability to think on one's own—the centerpiece of the first enlightenment—is not something we have lost, or even, as we shall see presently, something that biomimicry thinks we would be better off renouncing.

Yet a significant difference with Kant remains. Immaturity, Kant tells us, is self-incurred "if its cause is not lack of understanding, but lack of resolution and courage to use it without the guidance of another."[2] If we are to achieve enlightenment, we must have the courage and resolution not to turn to another for guidance. Not only, then, should we be *able* to think independently and autonomously, but we must also have the courage and resolution to *act on and realize* that ability. Over two hundred years later, however, it is becoming increasingly apparent that the seemingly courageous and resolute attempt to think independently and autonomously has in fact been extremely reckless; in the blink of an historical eye, it has brought us to the brink of an historic disaster. Humility, we may conclude, is now needed to recognize that guidance—specifically the guidance of nature—is needed after all.[3]

But how exactly are we to understand the guidance provided by nature? The original German word Kant uses in this context is *Leitung*, which may also be translated as "lead" (*Leitung* is also a cognate of the English "lead"). This suggests that to accept nature's guidance is to follow nature's lead. But, if there is certainly a sense in which biomimicry does call for us to turn to nature for guidance, this cannot involve thoughtless following, as though there were a set of transparent guidelines we could simply read out from the book of nature that would tell us how to live and what to do. Nature must be *interpreted* in a certain way, and whatever knowledge it contains must be appropriated *intelligently*, where intelligence is understood as a form of *legein*, and thus as involving careful selection and integration. There is, in other words, no mutually exclusive choice between either turning to nature for guidance, or thinking autonomously; the two may be combined through intelligent interpretation and appropriation of the guidance nature has to offer. Enlightened naturalism, one might surmise, is neither pure naturalism, nor pure enlightenment, but the intelligent gathering of the two.

It is also important to note that the "new way of viewing and valuing nature" Benyus proposes involves understanding nature not so much as a guide, or even as a teacher, but rather as a mentor. Throughout this book, I have referred to nature as both teacher and mentor without distinguishing between the two. To the extent that one learns from both teachers and mentors, this was justified. But there is nevertheless a significant difference between a teacher and a mentor. A teacher is someone who shows things to another who, lacking sufficient knowledge, also lacks autonomy. This, then, is why, when we think of a teacher, we typically think of an adult, and when we think of those whom they teach, we typically think of children.

A mentor, by contrast, provides guidance and advice to someone who, though they may possess sufficient knowledge to have achieved a measure of autonomy, lacks experience and wisdom. This guidance and advice may involve knowledge transmission, but this knowledge is provided in addition to, and as a means of refining or reorienting, a body of knowledge that has already been accumulated. If, for example, we knew little or nothing of genes, polymers, trophic cycles, and the like, then what we might learn from nature would be severely limited, just as it was for the ancients. This is why, when we think of a mentor, we think of a wise adult, and when we think of a mentee, we do not think of a child, but of a young adult, of someone who may know quite a bit already and may thus, at least in some respects—such as the use of the understanding and reason—be largely autonomous, but who, in their youth and inexperience, nevertheless lacks some of the knowledge, and certainly much of the experience and wisdom, of their mentor. Telemachus, to take the paradigmatic example, was no longer a child, yet he lacked the wisdom and experience of his father, Odysseus, to whom Mentor was but a friend, and could thus benefit from Mentor's greater reserves of knowledge, experience, and sagacity. Similarly, if Benyus today proposes we take nature as mentor, it is not, I suggest, because she is seeking to surrender our autonomy, but rather because, over the course of 3.8 billion years of evolution, nature has accumulated a wealth of knowledge, experience, and wisdom that we can ill afford to ignore.

This analysis in turn suggests that the first enlightenment—that of Kant—was not in fact the emergence into a state of full maturity, of genuine adulthood, but rather the self-affirmation of the adolescent who arrogantly thinks they can understand and accomplish everything on their own. It was, in other words, not the true attainment of maturity, but a stage in our history characterized by the hubristic belief that we need no guidance whatsoever, that our agriculture, industry, and cities, as well as the rules of conduct applicable therein, can all be free creations of ours, wholly independent of natural models and measures. Further, like an adolescent who, at the same time as they strive for autonomy, is living off of wealth set aside by their parents or elders, we have been seeking to realize our inner kernel of freedom while at the same time living off and destroying the wealth accumulated by nature. Indeed, it is only very belatedly that we are coming to realize that the autonomy we prematurely believed ourselves to have achieved has come at the cost of the unprecedented destruction of wealth accumulated by nature, and that, far from being genuinely autonomous, we remain highly dependent on nature for the myriad services it provides.

And yet, while we are increasingly aware of all the things nature *does for us*, of all the "ecosystem services" it provides, we remain largely ignorant of all the things nature may *show or teach us how to do*. Of course, showing or teaching us how to do things is, in a sense, also something nature does for us, an immaterial benefit that, as such, presumably belongs to the category of "cultural services."[4] In another sense, however, it is the very opposite of something nature does for us. When nature does things for us, we remain *dependent* on nature, but when nature shows us how to do things, we gain a measure of *autonomy* from nature. Further, what nature shows us how to do is, at least in part, how to reduce our dependence on her supporting, provisioning, and regulating services by learning how to accomplish at least some of these services ourselves. When we learn from nature how to recycle nutrients, we learn how to support ourselves. When we learn from nature how to generate and store solar energy, we learn how to provide for ourselves. And when we learn from nature how to reduce the urban heat island effect, we learn how to regulate local temperatures ourselves. If the ultimate goal of education is autonomy, then the same is true of the education we receive from nature.

This autonomy should not be confused, however, with radical independence, as if we could do everything that nature currently does for us entirely on our own. Just as the autonomy of an individual in society (e.g., their ability to provide for themselves, think for themselves) can only be realized through cooperating with others (e.g., employers, colleagues, friends), so the same is true of the autonomy of humanity in nature: it can only be realized through cooperating with, and living as a part of, nature. For the clearing we inhabit to be a clearing, there must be a surrounding forest, where by "forest" I mean more than woods, but, metonymically, that part of Gaia that lies beyond the human-built world and with which what goes on within the clearing must "fit" if we are to endure. The alternative, a world in which the surrounding forest has been destroyed, is not a clearing any longer, but a wasteland. And yet, in seeking to achieve radical autonomy from nature—in refusing any guide other than our own reason and our own senses—it is precisely a wasteland that we have inadvertently been bringing about.

In view of all of this, it is possible to chart the course of humankind through three different stages. First comes pre-enlightenment, the stage of childlike immaturity, in which human beings are unable to make full use of their understanding and reason, and instead defer to the guidance of another, especially by taking

God or His earthly representatives as mentor. Second comes the first enlightenment, a stage of liberation and self-affirmation characteristic of adolescence, in which humans throw off the shackles of dogmatic belief and instead learn to think for themselves, to use their intrinsic capacity for understanding and reasoning, in order to produce, act, and acquire knowledge on their own. And third comes what I am referring to as the "new enlightenment," the stage of early adulthood in which humans realize that their aspirations to autonomy were premature, that they remain dependent on another, on nature, and that, while the true path to autonomy requires us to leave unthinking following and obedience behind, this is not to say it requires us to affirm radical independence either. Rather, the path to autonomy lies in a willingness thoughtfully to learn from that other on whom we depend, to learn from nature.

These three stages may be charted at the four main levels of human existence: ontology, technics, ethics, and epistemology. At the level of ontology, the pre-enlightened stage finds humans believing their fate and destiny to lie in the hands of another, especially God. In the stage of the first enlightenment, humans believe that they hold their fate in their own hands, that a kernel of pure freedom means they may freely decide what they are to produce, do, and know. And, in the stage of the new enlightenment, they realize that there is no transcendental realm that exists both apart from and in parallel to nature, and that human freedom must instead be understood in the context of enlightened naturalism, and thus as a feature of those physei-beings that inhabit the clearing, itself understood as a space that has opened up within, but that is also constrained by, Gaia.

At the level of technics, the pre-enlightened stage finds humans practicing a form of imitation of nature that is both unthinking, in the sense of aiming for faithful reproduction, and largely devoid of understanding, in that the workings of nature are for the most part not well comprehended at all. In the stage of the first enlightenment, humans come to believe that they understand nature, that they have acquired much knowledge about nature, and that they may use that knowledge in order freely to create anything they wish—provided only that it conforms, as it must, to nature's universal laws. And, in the stage of the new enlightenment, humans come to realize that there is much that they may learn from nature, that a thoughtful imitation of nature, making extensive use of the human capacity to understand nature, ultimately enhances creativity, giving humans access to a much wider set of potential models—and, by implication, a much wider set of potential technological innovations—than they might have come up with on their own.

At the level of ethics, the pre-enlightened stage involves unthinking obedience to higher authorities, often religious. Ethics are encoded in sacred texts, and these texts tell humans what they must do. In the stage of the first enlightenment, humans learn what is right through independent thinking and reasoning, in particular through universalizing the maxims underlying their actions. And, in the stage of the new enlightenment, humans come to learn what is right by recognizing the self-evident good of another—the provision and maintenance of a habitable earth characteristic of Gaia—and then endeavoring to adopt that good as their own.

Finally, at the level of epistemology, the pre-enlightened stage finds humans simply accepting knowledge—or, at least, apparent instances of knowledge—handed down by others, especially religious authorities. In the stage of the first enlightenment, knowledge is acquired either on one's own, whether through the senses, reason, or introspection, or from other humans. And, finally, in the stage of the new enlightenment, humans come to recognize that much knowledge has already been accumulated in nature, that that knowledge has already been extensively tried and tested, and that, while they should not seek to reproduce it unthinkingly, for its abstraction and use must be carried out intelligently, there is nevertheless much scope not only for learning from it but also for doing so in a methodical and scientific manner.

But if the new philosophy set out in the present work may make an essential contribution to a new enlightenment, it is also true that it remains limited in scope, and that a new enlightenment would require much more than just this new philosophy. This limitation derives essentially from the fact that the new philosophy is only *environmental* in nature, that it is concerned with the being of nature and with the human relation to nature, but only tangentially with human nature and with the relation of human to fellow human. These latter issues were certainly touched upon—notably, when discussing the exposure of human beings to the clearing—but they were not directly addressed. The task of addressing these issues directly, and thus also of understanding the human and interhuman aspects of the new enlightenment, will, I hope, be the topic of a future work.

# Notes

## PREFACE

1. Freya Mathews, "Towards a Deeper Philosophy of Biomimicry," *Organization & Environment* 24, no. 4 (2011): 364–87.
2. Bernadette Bensaude-Vincent, "A Cultural Perspective on Biomimetics," in *Advances in Biomimetics*, ed. Anne George (InTech, 2011), https://www.intechopen.com/chapters/15671 (accessed July 20, 2022).
3. Vincent Blok and Bart Gremmen, "Ecological Innovation: Biomimicry as a New Way of Thinking and Acting Ecologically," *Journal of Agricultural and Environmental Ethics* 29, no. 2 (2016): 203–17.
4. Hub Zwart, "What Is Mimicked by Biomimicry? Synthetic Cells as Exemplifications of the Threefold Biomimicry Paradox," *Environmental Values* 28, no. 5 (2019): 527–49.
5. Aldo Leopold, "The Land Ethic," in *A Sand County Almanac* (Oxford: Oxford University Press, 1949), 209–10.
6. Janine Benyus, *Biomimicry: Innovation Inspired by Nature* (New York: Harper Perennial, 1997).
7. Henry Dicks and Vincent Blok, eds., "The Philosophy of Biomimicry and Bio-inspiration," special issue, *Environmental Values* 28, no. 5 (2019).

## INTRODUCTION

1. Janine Benyus, *Biomimicry: Innovation Inspired by Nature* (New York: Harper Perennial, 1997), 2.
2. Freya Mathews, "Towards a Deeper Philosophy of Biomimicry," *Organization & Environment* 24, no. 4 (2011): 364.
3. Bernadette Bensaude-Vincent, "A Cultural Perspective on Biomimetics," in *Advances in Biomimetics*, ed. Anne George (InTech, 2011), https://www.intechopen.com/chapters/15671 (accessed July 20, 2022); Vincent Blok and Bart Gremmen, "Ecological Innovation: Biomimicry as a New Way of Thinking and Acting Ecologically," *Journal of Agricultural and Environmental Ethics* 29, no. 2 (2016): 206–7.
4. Freya Mathews, "Can China Lead the World Towards an Ecological Civilization: A Manifesto," *Journal of Nanjing Forestry* 2 (2013): https://www.freyamathews.net/full-text-articles (accessed July 20, 2022).

5.    Benyus, *Biomimicry*, epigraph, n.p.

6.    See, for example, J. Vincent, O. A. Bogatyrev, N. R. Bogatyrev, A. Bowyer, and A. K. Pahl, "Biomimetics: Its Practice and Theory," *Journal of Colloid and Interface Science* 3, no. 9 (2006): 471–82.

7.    Wes Jackson, "Nature as Measure," in *Nature as Measure: The Selected Essays of Wes Jackson* (Berkeley: Counterpoint, [1994] 2011), 59–79.

8.    Benyus, *Biomimicry*, 11–58.

9.    B. Bensaude-Vincent, H. Arribart, Y. Bouligand, and C. Sanchez, "Chemists and the School of Nature," *New Journal of Chemistry* 26 (2002): 1–5.

10.   Bharat Bhushan, "Biomimetics: Lessons from Nature—an Overview," *Philosophical Transactions of the Royal Society A* 367 (2009): 1445–86.

11.   Jackson, "Nature as Measure," 185.

12.   Henry Dicks, "The Philosophy of Biomimicry," *Philosophy & Technology* 29, no. 3 (2016): 223–43.

13.   Immanuel Kant, *Critique of Pure Reason*, trans. Max Müller (London: Penguin, [1781, 1787] 2007), 256.

14.   Immanuel Kant, *Critique of Judgment*, trans. James C. Meredith (Oxford: Oxford University Press, 2007), 51–66.

15.   Kant, *Critique of Judgment*, 75–77.

16.   Discussions of technics are present also in the second critique, but only really by way of contrast to morality, in particular via the idea that technical imperatives are hypothetical rather than categorical.

17.   See Bryan G. Norton, "Conservation and Preservation: A Conceptual Rehabilitation," *Environmental Ethics* 8, no. 3 (1986): 190–220.

18.   See, for example, Holmes Rolston III, "Value in Nature and the Nature of Value," in *Philosophy and the Natural Environment*, ed. R. Attfield and A. Belsey (Cambridge: Cambridge University Press, 1994), 13–30.

19.   Perhaps the main exception is the strand of environmental philosophy that takes its inspiration from Heidegger. See, for example, Bruce Foltz, *Inhabiting the Earth: Heidegger, Environmental Ethics, and the Metaphysics of Nature* (New York: Humanity, 1995).

20.   See Christopher Preston, *Grounding Knowledge: Environmental Philosophy, Epistemology, and Place* (Athens: University of Georgia Press, 2003); and Jim Cheney, "Universal Consideration: An Epistemological Map of the Terrain," *Environmental Ethics* 20, no. 3 (1998): 265–77.

21.   Harper Douglas, "Etymology of Revolution," Online Etymology Dictionary, https://www .etymonline.com/search?q=revolution (accessed July 12, 2022).

22.   The idea of in some sense reviving early Greek thinking of nature overlaps in a number of important respects with Callicott's call for a "Neo-Presocratic revival," involving an articulation of the natural sciences with philosophy and the humanities. See J. Baird Callicott, "A Neo-Presocratic Manifesto," *Environmental Humanities* 2, no.1 (2013): 169–86.

23.   Martin Heidegger, *Introduction to Metaphysics*, trans. G. Fried and R. Polt (New Haven, CT: Yale University Press, 2000), 133.

24. See, for example, Levi Bryant, Nick Srnicek, and Graham Harman, eds., *The Speculative Turn: Continental Materialism and Realism* (Melbourne: re:Press, 2011); and Karen Barad, *Meeting the Universe Halfway: Quantum Physics and the Entanglement of Matter and Meaning* (Durham, NC: Duke University Press, 2007).

## 1. NATURE AS *PHYSIS*

1. Phillipe Descola, *Par-delà nature et culture* (Paris: Gallimard, 2005).

2. Bruno Latour, *Politiques de la nature: Comment faire entrer les sciences en démocratie* (Paris: Découverte, 1999), 42 (my translation).

3. Janine Benyus, *Biomimicry: Innovation Inspired by Nature* (New York: Harper Perennial, 1997), 7.

4. Jay Harman, *The Shark's Paintbrush: Biomimicry and How Nature Is Inspiring Innovation* (Ashland: White Cloud, 2013), 51-64.

5. Martin Heidegger, *The Fundamental Concepts of Metaphysics: World, Finitude, Solitude*, trans. W. McNeill and N. Walker (Indianapolis: Indiana University Press, 1995), 25.

6. Martin Heidegger, *Basic Problems of Phenomenology*, trans. A. Hofstadter (Indianapolis: Indiana University Press, 1988), 107.

7. Martin Heidegger, *The Principle of Reason*, trans. R. Lilly (Indianapolis: Indiana University Press, 1991), 63.

8. Martin Heidegger, "The Question Concerning Technology," *Basic Writings*, ed. D. F. Krell (Oxford: Routledge, 1993), 317 (my emphasis).

9. Heidegger, "Question Concerning Technology," 317.

10. Aristotle, *Physics*, 37.

11. See Heidegger, "The End of Philosophy and the Task of Thinking," in *Basic Writings*, 431-49.

12. Wikisource, "Fragments of Parmenides," https://en.wikisource.org/w/index.php?title=Fragments_of_Parmenides&oldid=10779923 (accessed April 25, 2022).

13. See Patricia Curd, *The Legacy of Parmenides* (Las Vegas: Parmenides, 2004), 75-80.

14. Anaximander, quoted in Martin Heidegger, "Anaximander's Saying," in *Off the Beaten Track*, trans. J. Young and K. Haynes (Cambridge: Cambridge University Press, 2002), 242.

15. For the canonical expression of this position, see G. E. L. Owen, *Logic, Science, and Dialectic: Collected Papers in Greek Philosophy* (Ithaca, NY: Cornell University Press, 1986).

16. Alexander Mourelatos, *The Route of Parmenides* (Las Vegas: Parmenides, 2008), 130-33.

17. Curd, *Legacy of Parmenides*.

18. Heidegger, "End of Philosophy," 433.

19. Kant himself was of course critical of the powers of pure reason to access "things in themselves."

20. Alfred N. Whitehead, *Process and Reality: An Essay in Cosmology* (New York: Free Press, 1978): 39.

21. Harper Douglas "Etymology of Science," Online Etymology Dictionary, https://www.etymonline.com/search?q=science (accessed July 12, 2022).

22. Harper Douglas "Etymology of Atom," Online Etymology Dictionary, https://www.etymonline.com/search?q=atom (accessed July 12, 2022).

23. David Hume, *An Enquiry Concerning Human Understanding* (Oxford: Oxford University Press, 2007), 16–17.

24. In Kant's own terms, the syntheses performed by the mind do not only bring together Ideas, for Kant reserves the term "Idea" for metaphysical "Ideas of reason," but also "representations" in general. If, however, the word "ideas" is used in the broader sense common to Kant's near contemporaries, such as Locke, then the various "syntheses" he posits do bring "ideas" together. On Kant's understanding of "Ideas," see Kant, *Critique of Pure Reason*, 297–302.

25. Nietzsche's genealogical studies, for example, show how certain ideas—notably the ideas of good and bad—change over time. See Friedrich Nietzsche, *On the Genealogy of Morality*, trans. M. Clark and A. J. Swensen (Indianapolis: Hackett, 1998).

26. Norbert Wiener, *Cybernetics: Or, Control and Communication in the Animal and the Machine* (New York: Technology, 1948); Norbert Wiener, *The Human Use of Human Beings: Cybernetics and Society* (Boston: Da Capo, [1950] 1954).

27. Heidegger, "End of Philosophy."

28. Heidegger does not say this directly. What he says is that poiēsis is concealed within the roots of technology ("Question Concerning Technology," 334–35). But, given that physis is a mode of poiēsis, and given that he also thinks technology is cybernetic in character (Heidegger, "End of Philosophy," 434), we may infer that he thought physis was concealed within the roots of cybernetics.

29. Humberto Maturana and Francisco Varela, *Autopoiesis and Cognition: The Realization of the Living* (Dordrecht: D. Reidel, 1980).

30. Maturana, "Introduction," in *Autopoiesis and Cognition*, xvii.

31. Martin Heidegger, *Introduction to Metaphysics*, trans. G. Fried and R. Polt (New Haven, CT: Yale University Press, 2000), 15.

32. What one might call the "ontological blindness" of Maturana's understanding of autopoiesis is visible when, addressing the relation between autopoiesis and ontology, he adopts a standard, post-Parmenidean conception of being as "permanent enduring" and then goes on to identify autopoiesis not with "being," but with "doing," as if being and doing were necessarily opposed. See Humberto Maturana and Bernhard Poulsen, *From Being to Doing: The Origins of the Biology of Cognition* (Heidelberg: Carl Auer, 2004).

33. Edgar Morin, *La méthode, tome 1: La nature de la nature* (Paris: Seuil, 1977).

34. There can be little doubt that Maturana and Varela's concept of autopoiesis also exerted a significant influence on Morin's conceptualization of nature as self-production.

35. Morin, *La méthode*, 182–235.

36. Morin, 277–80.

37. Morin, 183–84, 231.

38. Edgar Morin, *Le paradigme perdu: la nature humaine* (Paris: Seuil, 1973).

39. Niklas Luhmann, *Social Systems* (Stanford, CA: Stanford University Press, 1995).

40. Morin, *La méthode*, 27.

41. Morin, 235.

42. Morin, 157 (my translation).

43. Morin, 158.

44.  Heidegger, "Question Concerning Technology," 326.
45.  Martin Heidegger, "On the Essence and Concept of *Physis* in Aristotle's *Physics B*, 1," in *Pathmarks*, trans. T. Sheehan (Cambridge: Cambridge University Press, 1998), 222.
46.  Heidegger, "End of Philosophy," 442.
47.  This understanding of the limit or boundary corresponds to the Greek concept of the *peras*, which Heidegger understands in terms of "self-delimitation," as well as the "Being of beings": "The *peras* is not something that accrues to a being from outside. Much less is it some deficiency in the sense of a detrimental restriction. Instead, the self-restraining hold that comes from a limit . . . is the Being of beings; it is what first makes a being be a being as opposed to a nonbeing. For something to take such a stand therefore means for it to attain its limit, to de-limit itself." Heidegger, *Introduction to Metaphysics*, 63. In a similar vein, Heidegger also remarks that "beings are for them [i.e., the Greeks] the self-limiting, over and against the limitless and dissolving." Martin Heidegger, *Basic Questions of Philosophy: Selected "Problems" of "Logic,"* trans. R. Rojcewicz and André Schuwer (Indianapolis: Indiana University Press, 1994), 119.
48.  Harper Douglas "Etymology of Existence," Online Etymology Dictionary, https://www.etymonline.com/search?q=existence (accessed July 12, 2022).
49.  This claim requires qualification. Science is not only concerned with cutting things up into their constituent elements but also with how these elements fit together and interact. This latter concern means that scientists will also frequently construct or assemble things, but these constructions differ from those of technē in that they are carried out primarily with a view to understanding nature, rather than with a view to making entities that serve some purpose or other.
50.  Karen Barad also theorizes science as a "cutting" process. Against the traditional view of elementary particles *inter-acting* with one another, Barad argues, first, that what is taking place are *intra-actions* of "entangled matter," and, second, that the role of science is to enact cuts separating entangled matter out into different things, including the all-important cut made between human subjects and natural objects. See Karen Barad, *Meeting the Universe Halfway: Quantum Physics and the Entanglement of Matter and Meaning* (Durham, NC: Duke University Press, 2007). What Barad fails to see, however, is that, in producing themselves certain entities *cut themselves out*, thus creating a division between internal intra-actions, and external inter-actions with the environment.
51.  It is instructive to compare the concept of "ensistence" with Kant's concept of the "thing in itself," or "noumenon." The reason Kant thinks that the noumenon does not appear is that it remains *in itself* and, as such, is not ex-posed to humans. By contrast, the reason that which ensists does not appear, the reason that it is not ex-posed to humans, is that it is *held within some other reality*. Or, to put it another way, whereas the noumenon does not appear because it is *forever cut off* from human beings, that which ensists does not appear because it has *yet to be cut out* by human beings. Further, whereas the fact that the noumena are cut off from the synthesizing capacities of the human mind means they are also radically separate from one another, existing not in some greater reality, but only in themselves, the opposite is true of beings that ensist. The latter are not "in themselves," in the sense of being immanent to themselves, for what they are immanent to or held within is rather some other reality; it is, in other words, their *absence of separation* from other entities that means that they do not appear, whereas in the case of the

noumena it is their *total separation* from other entities which prevents them from appearing.

52.    In the narrow sense in which Heidegger uses the word "philosophy" in "The End of Philosophy," this word refers to metaphysics and, as such, to something that only came into existence with Plato. For my part, I use the word "philosophy" in a wider sense, such that it refers also to the thinking of physis undertaken first by the ancient Greeks and then again by a new generation of thinkers, including Heidegger himself. When speaking of philosophy in Heidegger's narrow sense, then, I will usually use the word "metaphysics." As for the expressions "thinking" or "thought," I use them to refer to philosophy in the wide sense, but also, in the wake of Parmenides, to that which split into atomism and idealism, into science and philosophy (in the narrow sense).

53.    Maturana and Varela are obvious examples of this.

54.    A key historical example of this way of thinking, and that incidentally shows that it is not restricted to science (for it is present in post-Parmenidean thinking in general), comes from Duns Scotus. Noting that a thing may be produced, Duns Scotus then observes that it may, at least hypothetically, be produced by nothing, by itself, or by something else. He then rules out the possibility that it may be produced by nothing, for "that which is nothing is cause of nothing," and he also rules out the possibility that it may be produced by itself, for "there is nothing that makes or generates itself." The only remaining possibility, then—and also the only one retained by science—is that every entity is produced by some other entity. See Duns Scotus, *Ordinatio* 1.2.43, http://www.logicmuseum.com/wiki/Authors/Duns_Scotus/Ordinatio/Ordinatio_I/D2/Q2B (accessed June 21, 2022).

55.    On this topic, see Edgar Morin, *La méthode, tome 2: La vie de la vie* (Paris: Seuil, 1980), 139.

56.    Kant, *Critique of Pure Reason*, 291–96.

57.    Harper Douglas "Etymology of Articulate," Online Etymology Dictionary, https://www.etymonline.com/search?q=articulate (accessed July 12, 2022).

58.    Morin, *La méthode*, 94.

59.    Morin, 225.

60.    Morin, 98.

61.    Richard Dawkins, *The Selfish Gene* (Oxford: Oxford University Press, [1976] 1989), 36.

62.    Dawkins, *Selfish Gene*, v.

63.    Dawkins, 46–52.

64.    Stuart Kauffman, *At Home in the Universe: The Search for the Laws of Self-Organization and Complexity* (Oxford: Oxford University Press, 1995), 49.

65.    For an exposition of this latter theory, see Freeman Dyson, *Origins of Life*, 2nd ed. (Cambridge: Cambridge University Press, 1999).

66.    Kauffman, *At Home in the Universe*, 73.

67.    Kauffman, 66.

68.    Maturana, "Introduction," in *Autopoiesis and Cognition*, 12.

69.    Warwick Fox, *Toward a Transpersonal Ecology: Developing New Foundations for Environmentalism* (Totnes: Resurgence, [1990] 1995), 172.

70.    See, for example, Heidegger, *Introduction to Metaphysics*, 131–32.

71.    See Peter Miller, *Smart Swarm* (London: Harper Collins, 2010).

72.  Andrew M. Simons, "Many Wrongs: The Advantage of Group Navigation," *Trends in Ecology and Evolution* 19, no. 9 (2004): 453–55; E. A. Codling, J. W. Pitchford, and S. D. Simpson, "Group Navigation and the 'Many Wrongs Principle' in Models of Animal Movement," *Ecology* 88, no. 7 (2007): 1864–70.

73.  Maturana, "Biology of Cognition," 11.

74.  Aldo Leopold, "The Land Ethic," in *A Sand County Almanac* (Oxford: Oxford University Press, 1949), 204.

75.  Henry A. Gleason, "The Individualistic Concept of the Plant Association," *Bulletin of the Torrey Botanical Club* 53, no. 1 (1926): 7–26.

76.  On this topic, see also Fritjof Capra, *The Web of Life* (London: Harper Collins, 1997), 208–10.

77.  James Lovelock, *Gaia: A New Look at Life on Earth* (Oxford: Oxford University Press, [1979] 2009), 44–58.

78.  Janine Benyus, quoted in Daniel C. Wahl, "Learning from Nature and Designing as Nature: Regenerative Cultures Create Conditions Conducive to Life," Biomimicry Institute, September 6, 2016, https://biomimicry.org/learning-nature-designing-nature -regenerative-cultures-create-conditions-conducive-life/?msclkid=98d41688b4d411eca 45905fd31d0a95e.

79.  Lovelock, *Gaia*, 30–43.

80.  One could perhaps wonder why Benyus talks of the conditions created by life as "conducive to" life rather than "productive of" life, as would appear to be required by the concept of Gaian self-production. One plausible answer is that, since living beings produce themselves, the conditions they create for themselves are only conducive to their existence and not productive of their existence, for the latter would imply that living beings are produced by things other than themselves. But self-producing entities can also participate in higher-order self-producing entities, as occurs also when individuals participate in societies or species, and at these higher levels what appears at a lower level as merely conducive to the existing of a self-producing being is productive of that being, viewed now as a part in a higher-order self-producing entity or system.

81.  See P. F. Hoffman, A. J. Kaufman, G. P. Halverson, and D. P. Schrag, "A Neoproterozoic Snowball Earth," *Science* 281 (1998): 1342–46.

82.  Barry Commoner, *The Closing Circle: Nature, Man, and Technology* (New York: Knopf, 1971), 299.

83.  See Yves Christen, *L'animal est-il une personne?* (Paris: Flammarion, 2011), 49–89.

84.  Heidegger, "End of Philosophy," 442.

85.  Heidegger, 443.

86.  Daniel Dennett, for example, has claimed that consciousness is an illusion created by the brain. See Daniel Dennett, *Consciousness Explained* (Boston: Little, Brown, 1981).

87.  This dualist position is associated above all with Descartes.

88.  Mark Rowlands, for example, contends that the operation of the mind extends outside the head and into the external world. See Mark Rowlands, "The Extended Mind," *Zygon* 44, no. 3 (2009): 628–41.

89.  The use of the Kantian expression "things in themselves" should not be taken to imply that the entities in question are necessarily conceived in a Kantian manner, and thus as

outside of space and time, for space and time may be—and usually are—assumed to be intrinsic features or properties of the subject-independent physical realm.

90.　Maturana and Varela, for example, are aware only of the internal space of biological cognition and the external space of the living being's environment, which, they think, is necessarily inaccessible, even if an observer may in some cases see the two as "structurally coupled." But the idea that there may be some "third space," the clearing, in which there appear neither subjective representations nor things in themselves, is not considered.

91.　Heidegger, *Basic Questions of Philosophy*, 23.

92.　This preontological understanding of being may be contrasted with the ontological understanding of being, which involves taking the being and existence of things as the explicit object of thought, rather than, as in the case of the preontological understanding, as understood but without being explicitly thought about.

93.　See Martin Heidegger, *Being and Time*, trans. John Macquarrie and Edward Robinson (Oxford: Blackwell, [1927] 1995), 149-68.

94.　Heidegger famously avoided talking of "humans" in *Being and Time*, preferring instead to talk of *Dasein*, a word I interpret as referring to any being that is exposed to the clearing. But, while Heidegger's approach has the important consequence of allowing philosophical reflection into that being that understands being to stand apart from and prior to anthropology, understood as an empirical science, provided we see the distinctive feature of humans as their exposure to the clearing, I see no problem in using the word "humans" in this context, rather than *Dasein*.

95.　Heidegger, *Being and Time*, 156-57.

96.　Heidegger, 168.

97.　Michael Tomasello, *Origins of Human Communication* (Cambridge, MA: MIT Press, 2010), 3.

98.　Tomasello, *Origins of Human Communication*, 34-37.

99.　Tomasello, 38.

100.　Tomasello, 116-25.

101.　According to Heidegger, the same is true also of ancient Greek thinking of physis: "A historical reflection will realize that the Greek theory of natural processes did not rest on insufficient observation but on another—perhaps even deeper—conception of nature that precedes all particular observations." Heidegger, *Basic Questions of Philosophy*, 47-48.

102.　In the English language, the conceptual link between the words "clearing" and "enlightened" is somewhat faint, for it is not immediately apparent on the basis of the words alone. In French, by contrast, the link is much more obvious. The French word for "enlightened" is *éclairé*. And the French word for "clearing"— which Heidegger incidentally claims is the original word of which the German word for clearing, *Lichtung*, is but a translation (Heidegger, "End of Philosophy," 441)— is *clairière*. In the French language, much more so than in English, there thus resonates the conceptual proximity between the adjective "enlightened" (*éclairé*) and the noun "clearing" (*clairière*); the nature theorized by enlightened naturalism is a nature that appears in the *clairière* (clearing) and, as such, is *éclairée* (enlightened).

103.　In keeping with this, when postmodern theorist Karen Barad discusses biomimicry, she contests the sharp nature/culture divide on which she thinks biomimicry is based. See Barad, *Meeting the Universe Halfway*, 367-68.

104. In a world no longer split radically between subjects and objects, there thus emerge all sorts of hybrids between the two, such as Latour's "quasi-subjects" and "quasi-objects." See Bruno Latour, *Nous n'avons jamais été modernes* (Paris: Découverte, 1991), 105.

105. The word "nature," it is worth noting, is not the only one common to the principles of nature as model, nature as measure, and nature as model, for the word "as" is also present in each of these principles, in which case to provide a full account of these principles we also need a theory of the "as"— something that enlightened naturalism can provide.

106. This allows us to see another contrast with modern dualism. Whereas a basic problem for modern dualism is how to understand the belonging of the mind to the body, the equivalent problem for enlightened naturalism is how to understand the belonging of the clearing to Gaia. Further, however exactly enlightened naturalism answers this question, it is clear that it takes on a distinctly ecological dimension—entirely lacking from modern dualism with its focus on the individual—that may in turn underpin a radically new ethics.

## 2. NATURE AS MODEL

1. Martin Heidegger, *Basic Questions of Philosophy: Selected "Problems" of "Logic,"* trans. R. Rojcewicz and André Schuwer (Indianapolis: Indiana University Press, 1994), 35–39.

2. See, for example, M. Franssen, G. J. Lokhorst, and I. van de Poel, "Philosophy of Technology," in *The Stanford Encyclopedia of Philosophy*, ed. Edward N. Zalta, http://plato .stanford.edu/archives/fall2015/entries/technology/ (accessed June 23, 2022).

3. This is not to say that I will, in what follows, also be developing a theory of the imitation of nature in art, understood as something distinct from technology. It is certainly possible, however, that the theory of biomimetic technics presented in this chapter could contribute to that project.

4. Plato, *The Republic*, trans. Desmond Lee (London: Penguin Classics, 1974), 363.

5. Plato, *Republic*, 362.

6. See Hans Blumenberg, "Imitation of Nature: Toward a Prehistory of the Idea of the Creative Being," *Qui parle?* 12, no. 1 (2000): 28–29.

7. Blumenberg, "Imitation of Nature," 29.

8. K. Freeman, *Ancilla to the Pre-Socratic Philosophers (A Complete Translation of the Fragments in Diels, Fragmente der Vorsokratiker)* (Cambridge, MA: Harvard University Press, 1948), 154.

9. Aquinas, quoted in Ananda Coomaraswamy, "The Nature of Medieval Art," *Studies in Comparative Religion* 15, nos. 1–2 (1983), http://www.studiesincomparativereligion.com /public/articles/The_Nature_of_Medieval_Art-by_Ananda_K_Coomaraswamy.aspx (accessed July 15, 2022).

10. Nicholas of Cusa, "Idiota da Mente," in *Nicholas of Cusa on Wisdom and Knowledge*, trans. Jasper Hopkins (Minneapolis: Arthur J. Banning, 1996), 538.

11. Françoise Choay, "La ville et le domaine bâti comme corps dans les textes des architectes-théoriciens de la première renaissance italienne," *Nouvelle revue de psychanalyse* 9 (1974): 239–51.

12. Choay, "La ville et le domaine bâti," 247.

13. Thomas Hobbes, "Selections from the *Leviathan*," in *Philosophers Speak for Themselves: from Descartes to Locke*, ed. T. Smith and M. Grene (Chicago: University of Chicago Press, 1967), 157.

14. Hobbes, "Selections from the *Leviathan*," 157.

15. G. W. F. Hegel, *Hegel's Aesthetics, Lectures on Fine Art*, vol. 1, trans. T. M. Knox (Oxford: Clarendon, 1975), 41–42.

16. Hegel, *Hegel's Aesthetics*, 42.

17. Hegel, 45.

18. Hegel, 43.

19. Ernst Kapp, *Elements of a Philosophy of Technology: On the Evolutionary History of Culture*, trans. L. K. Wolfe (Minneapolis: University of Minnesota Press), 2018.

20. Kapp, *Elements of a Philosophy of Technology*, 27–33.

21. Max Eyth, *Lebendige Kräfte: Sieben Vorträge aus dem Gebiete der Technik*, 4th ed., (Berlin: J. Springer, 1924), 1, my translation.

22. Friedrich Dessauer, "Technology in Its Proper Sphere," in *Philosophy and Technology: Readings in the Philosophical Problems of Technology*, 2nd ed., ed. Karl Mitcham and Robert Mackey (New York: Free Press, 1983), 334.

23. Dessauer, "Technology in Its Proper Sphere," 320–21.

24. Friedrich Dessauer. *Philosophie der Technik. Das Problem der Realisierung* (Bonn: Cohen, 1927), 150 (my translation).

25. Ernst Cassirer, "Form and Technology," in *The Warburg Years (1919–1933): Essays on Language, Art, Myth, and Technology*, trans. S. G. Loft and A. Calcagno (New Haven, CT: Yale University Press, 2014), 302.

26. Blumenberg, "Imitation of Nature," 24.

27. Blumenberg, 46.

28. Martin Heidegger, "The Question Concerning Technology," *Basic Writings*, ed. D. F. Krell (Oxford: Routledge, 1993), 314–16.

29. Since Dessauer focused on *modern* technology, this example is anachronistic and inappropriate. An updated example might be, say, the problem of how to mass produce steel saucepans.

30. Charles Jencks, *The Story of Postmodernism* (Chichester: John Wiley, 2011).

31. Hub Zwart, "What Is Mimicked by Biomimicry? Synthetic Cells as Exemplifications of the Threefold Biomimicry Paradox," *Environmental Values* 28, no. 5 (2019): 527–49.

32. See, for example, Francisco Varela and Paul Bourgine, eds., *Toward a Practice of Autonomous Systems: Proceedings of the First European Conference on Artificial Life* (Cambridge, MA: MIT Press, 1992).

33. Lisa Grossman, "Polystyrene Atoms Could Surpass the Real Deal," New Scientist, October 31, 2012, https://www.newscientist.com/article/dn22440-polystyrene-atoms-could-surpass-the-real-deal/.

34. ISO 18458, "Biomimetics—Terminology, Concepts, and Methodology," 2015, https://www.iso.org/obp/ui/#iso:std:iso:18458:ed-1:v1:en (accessed June 23, 2022).

35. ISO 18458, "Biomimetics."

36. ISO 18458.

37. Tom McKeag, "Auspicious Forms: Designing the Sanyo Shinkansen 500-Series Bullet Train," *Zygote Quarterly* 2 (2012): 14–35.

38. It may seem contradictory to talk here of a "natural" system that has been "artificially delimited." In the present context, I use the word "natural" in the restrictive sense of a system that has not been *constructed* by humans, for it belongs to a being that constructs itself, but that has nevertheless been *delimited* by humans. Strictly speaking, then, such a system is partly natural, and partly artificial, though it would be laborious and confusing to try to reflect this hybrid status, and I shall thus opt for the simpler solution of talking of "natural systems."

39. See, for example, P. Gruber, D. Bruckner, H.-B. Schmiedmayer, H. Stachelberger, I. C. Gebeschuber, *Biomimetics—Materials, Structures, Processes* (Berlin: Springer Verlag, 2011); and Julian Vincent, "Biomimetics—A Review," *Journal of Engineering in Medicine* 223, no. 8 (2009): 919-39.

40. For a comprehensive discussion of Aristotle's theory of abstraction, see Alan Bäck, *Aristotle's Theory of Abstraction* (Dordrecht: Springer, 2014).

41. Humberto Maturana and Francisco Varela, *Autopoiesis and Cognition: The Realization of the Living* (Dordrecht: D. Reidel, 1980).

42. Freya Mathews, "Towards a Deeper Philosophy of Biomimicry," *Organization & Environment* 24, no. 4 (2011): 368.

43. Maturana, it follows, is not wrong to identify autopoiesis with doing. His error is to identify it with doing *and not with being*, for what self-producing beings are (i.e., self-producing) is also what they do (produce themselves).

44. Aristotle, *Physics*, trans. Robin Waterfield (Oxford: Oxford University Press, 2000), 38.

45. McKeag, "Auspicious Forms," 26-30.

46. B. Karthick and R. Maheshwari, "Lotus-Inspired Nanotechnology Applications," *Resonance* 13, no. 12 (2008): 1141-45.

47. This is not to deny, of course, that the morphology of geological features changes over time, but for purposes of human design its relative stability allows us to treat it as a "structure."

48. Frederick Ward, *Discours prononcé à la séance d'ouverture du Congrès International de Bienfaisance* (Brussels: Librairie Européenne, C. Muquardt, 1856), 9-31.

49. D. Dasgupta, ed., *Artificial Immune Systems and Their Applications* (Berlin: Springer, 1999).

50. Y. H. Jung, B. Park, J. U. Kim, and T. Kim, "Bioinspired Electronics for Artificial Sensory Systems," *Advanced Materials* 31 (2019), doi:10.1002/adma.201803637 (accessed July 15, 2022).

51. Lazaros S. Iliadis, Vera Kurkova, and Barbara Hammer, "Brain-Inspired Computing and Machine Learning," *Neural Computing & Applications* 32 (2020): 6641-43.

52. Ranil Senanayake and John Jack, *Analogue Forestry: An Introduction* (Clayton: Department of Geography and Environmental Science, Monash University, 1998).

53. See, for example, K. J. Young, "Mimicking Nature: A Review of Successional Agroforestry Systems as an Analogue to Natural Regeneration of Secondary Forest Stands," in *Integrating Landscapes: Agroforestry for Biodiversity Conservation and Food Sovereignty. Advances in Agroforestry,* vol. 12, ed. F. Montagnini (Cham: Springer, 2017), 179-209.

54. H. Dicks, J.-P. Bertrand-Krajewski, C. Ménézo, Y. Rahbé, J.-L. Pierron, and C. Harpet, "Applying Biomimicry to Cities: The Forest as Model for the City," in *Technology and the City: Towards a Philosophy of Urban Technologies,* ed. M. Nagenborg, T. Stone, M. G. Woge, and P. E. Vermass (Cham: Springer, 2020), 271-88.

55. Wes Jackson, *New Roots for Agriculture* (Lincoln: University of Nebraska Press, 1980).

56. Nancy Jack Todd and John Todd, *From Ecocities to Living Machines: Principles of Ecological Design* (Berkeley: North Atlantic, [1984] 1993).

57. Thierry Lefèvre and Michèle Auger, "Spider Silk as a Blueprint for Greener Materials: A Review," *International Materials Reviews* 61, no. 2 (2016): 127–53.

58. Shruthy Kuttappan, Dennis Mathew, and Manitha B. Nair, "Biomimetic Composite Scaffolds Containing Bioceramics and Collagen/Gelatin for Bone Tissue Engineering—a Mini Review," *International Journal of Biological Macromolecules* 93, pt. B (2016): 1390–401.

59. Michelle Oyen, "Dreaming Big with Biomimetics: Could Future Buildings Be Made with Bone and Eggshells?" The Conversation, March 8, 2016, https://theconversation.com /dreaming-big-with-biomimetics-could-future-buildings-be-made-with-bone-and -eggshells-55739.

60. Benyus, *Biomimicry*, 118–29.

61. G. Du, A. Mao, J. Yu, J. Hou, N. Zhao, J. Han, Q. Zhao, W. Gao, T. Xie, and H. Bai, "Nacre-Mimetic Composite with Intrinsic Self-Healing and Shape-Programming Capability," *Nature Communications* 10 (2019), doi: 10.1038/s41467-019-08643-x (accessed July 15, 2022).

62. J. Vincent, O. A. Bogatyrev, N. R. Bogatyrev, A. Bowyer, and A. K. Pahl, *Biomimetics: Its Theory and Practice, J. R. Soc. Interface* 3 (2006): 477–78.

63. Parvez Alam, "Biomimetic Composite Materials Inspired by Wood," in *Wood Composites*, ed. Martin P. Ansell (Cambridge: Woodhead, 2015), 357–394.

64. "Nature's Unifying Patterns," Biomimicry Institute, https://toolbox.biomimicry.org/core -concepts/natures-unifying-patterns/chemistry/ (accessed April 22, 2022).

65. Heungsoo Shin, Seongbong Jo, and Antonios G. Mikos, "Biomimetic Materials for Tissue Engineering," *Biomaterials* 24, no. 24 (2003): 4353–64.

66. See Freya Mathews, "Biomimicry and the Problem of Praxis," *Environmental Values* 28, no. 5 (2019): 573–99.

67. Stephen Mann, "Biomineralization and Biomimetic Materials Chemistry," *Journal of Materials Chemistry* 5 (1995): 935–46.

68. Benyus, *Biomimicry*, 97.

69. Y. Yang, X. Song, X. Li, Z. Chen, Q. Zhou, and Y. Chen, "Recent Progress in Biomimetic Additive Manufacturing Technology: From Materials to Functional Structures," *Advanced Materials* 30 (2018), doi: 10.1002/adma.201706539 (accessed July 15, 2022).

70. Jean-Claude Ameisen, *La sculpture du vivant: Le suicide cellulaire et la mort créatrice* (Paris: Seuil, 2003).

71. Michael S. Packer and David R. Liu, "Methods for the Directed Evolution of Proteins," *Nature Review Genetics* 16 (2015): 379–94.

72. Senanayake and Jack, *Analogue Forestry*.

73. A lots of littles approach could potentially lead production to be placed in the hands of the people, as Karl Marx famously desired, rather than in the hands of large companies or the state, and this could lead in turn to much more egalitarian societies.

74. Benyus, *Biomimicry*, 7.

75. Stephen J. Gould and Elizabeth S. Vrba, "Exaptation—a Missing Term in the Science of Form," *Paleobiology* 8, no. 1 (1982): 4–15.

76.    Michael Braungart and William McDonough, *Cradle to Cradle: Re-Making the Way We Make Things* (London: Vintage, 2009), 138.

77.    E. Kostal, S. Stroj, S. Kasemann, M. Matylitsky, and M. Domke, "Fabrication of Biomimetic Fog-Collecting Superhydrophilic–Superhydrophobic Surface Micropatterns Using Femtosecond Lasers," *Langmuir* 34, no. 9 (2018): 2933–41.

78.    Stěpanká Kadochová and Jan Frouz, "Thermoregulation Strategies in Ants in Comparison to Other Social Insects, with a Focus on Red Wood Ants (*Formica rufa* Group)," *F1000Research* 2 (2014), doi: 10.12688/f1000research.2-280.v2 (accessed July 15, 2022).

79.    Aristotle, *Physics*, 37, 51.

80.    René Girard, *Des choses cachées depuis la fondation du monde* (Paris: Grasset et Fasquelle, 1978), 17.

81.    Girard, *Des choses cachées*, 17–18.

82.    Blumenberg, "Imitation of Nature."

83.    "Smartflower," https://smartflower.com/ (accessed April 22, 2022).

84.    Wikisource, "The *Poetics* Translated by S. H. Butcher/1," https://en.wikisource.org/w/index.php?title=The_Poetics_translated_by_S._H._Butcher/1&oldid=12022729 (accessed April 22, 2022).

85.    Julian Vincent et al., "Biomimetics," 474.

86.    As a general rule, these other reasons will arise from the need or desire for the entity to fulfil different or additional functions.

87.    Yunhong Liu and Guangji Li, "A New Method for Producing 'Lotus Effect' on a Biomimetic Shark Skin," *Journal of Colloid and Interface Science* 388, no. 1 (2012): 235–42.

88.    Observations of birds in flight by Otto Lilienthal and the Wright brothers are the most famous moments in the history of understanding and imitating bird flight.

89.    Paul Ricoeur, *Temps et récit: tome 1, l'intrigue et le récit historique* (Paris: Seuil, 1983), 93.

90.    Blumenberg, "Imitation of Nature," 46.

91.    See, for example, Joe Kaplinsky, "Biomimicry Versus Humanism," *Architectural Design* 76, no. 1 (2006): 66–71.

92.    See P. E. Fayemi, K. Wanieck, C. Zollgrank, M. Maranzan, and A. Aoussat, "Biomimetics: Process, Tools, Practice," *Bioinspiration and Biomimetics* 12 (2017), doi: 10.1088/1748-3190/12/1/011002 (accessed July 15, 2022).

93.    For an overview of eyes as sources of inspiration for cameras, see G. J. Lee, C. Choi, D. H. Kim, and Y. M. Song, "Bioinspired Artificial Eyes: Optic Components, Digital Cameras, and Visual Prostheses," *Advanced Functional Materials* 28, no. 24 (2017), doi: 10.1002/adfm.201705202 (accessed July 15, 2022).

94.    Kapp, *Elements of a Philosophy of Technology*, 7–8.

95.    Martin Heidegger, *The Principle of Reason*, trans. R. Lilly (Indianapolis: Indiana University Press, 1991), 64.

96.    As Aristotle himself remarks: "The simile also is a metaphor; for there is very little difference. When the poet says of Achilles, "he rushed on like a lion," it is a simile; if he says, "a lion, he rushed on," it is a metaphor; for because both are courageous, he transfers the sense and calls Achilles a lion. . . . Similes must be used like metaphors, which only differ in the manner stated." Wikisource, "*Rhetoric* (Freese)/Book 3," https://en.wikisource.org/w/index.php?title=Rhetoric_(Freese)/Book_3&oldid=3515996) (accessed April 25, 2022).

97.    Max Black, "Metaphor," *Proceedings of the Aristotelian Society* 55 (1954–1955): 283.

98.    Black, "Metaphor," 279.

99.    Black, 285.

100.   For these objections, as well as Black's response, see Max Black, "More About Metaphor," *Dialectica* 31, no. 34 (1977): 431–57.

101.   Black makes a very similar point when he writes: "Since we must necessarily read 'behind the words,' we cannot set firm bounds to the admissible interpretations: ambiguity is a necessary by-product of the metaphor's suggestiveness." Black, "More About Metaphor," 444.

102.   Braungart and McDonough, *Cradle to Cradle*, 139.

103.   It is worth nothing that in the biomimicry literature it is certainly not uncommon to use the textbook form of metaphor to express essentially the same idea (i.e., imagining similarities between the antecedently dissimilar) as Braungart and McDonough express in the textbook form of simile. An example is Jerry Brunetti, *The Farm as Ecosystem: Tapping Nature's Reservoir—Geology, Biology, Diversity* (Greeley, CO: Acres USA, 2013).

104.   For these and other possible transfers from trees to buildings, see Dicks et al., "Applying Biomimicry to Cities."

105.   Black makes a very similar point when he remarks that the two subjects of metaphor "are often best regarded as 'systems of things,' rather than 'things.'" Black, "Metaphor," 291.

106.   Black, 293.

107.   On polar bears as providing models for architecture, see Hugh Aldersey-Williams, "Towards Biomimetic Architecture," *Nature Materials* 3 (2004): 277-79.

108.   Albert Howard, *An Agricultural Testament* (Mapusa: Other India, 1956).

109.   See Henry Dicks, "From Anthropomimetic to Biomimetic Cities: The Place of Humans in 'Cities like Forests,'" *Architecture Philosophy* 3, no. 1 (2018): 65–68.

110.   Howard, *Agricultural Testament*, 223.

111.   Kapp, *Elements of a Philosophy of Technology*, 172-77.

112.   See Dicks, "From Anthropomimetic to Biomimetic Cities," 98-99.

113.   There has been some discussion of biomimetic analogy, but it would not appear to have been recognized as a specific innovation scenario different, as such, from the widely recognized ones of biological push and technological pull. See, for example, K. Fu, D. Moreno, M. Yang, and K. L. Wood, "Bio-Inspired Design: An Overview Investigating Open Questions from the Broader Field of Design-by-Analogy," *Journal of Mechanical Design* 136, no. 11 (2014), doi: 10.1115/1.4028289 (accessed July 15, 2022); and J. M. O'Rourke and C. C. Seepersad, "Toward a Methodology for Systematically Generating Energy-and Materials-Efficient Concepts Using Biological Analogies," *Journal of Mechanical Design* 137, no. 9 (2015), doi:10.1115/1.4030877 (accessed July 15, 2022). As for the notion of biomimetic metaphor, as far as I'm aware it has not yet received any sustained theoretical attention or been discussed as a specific innovation scenario.

114.   Gilbert Simondon, *Du mode d'existence des objets techniques* (Paris: Aubier, 1989), 56-57.

115.   This is particularly common in the literature on smart cities, which often presents the analogy of the sentient organism as foundational. Antoine Picon, for example, argues that cities should be made "smart" in a much more "literal" acceptation than usual—that is to say, through the creation of a sentient urban organism capable of learning,

understanding, and reasoning. See Antoine Picon, *Smart cities: Théorie et critique d'un idéal auto-réalisateur* (Paris: Collection Actualités, 2013): 9, 23.

116.  See K. H. Jeong, J. Kim, and L. P. Lee, "Biologically Inspired Artificial Compound Eyes," *Science* 312, no. 5773 (2006): 557–61.

117.  Julian Vincent and Darrell Mann, "Systematic Technology Transfer from Biology to Engineering," *Philosophical Transactions of the Royal Society A* 360, no. 1791 (2002): 159–73.

118.  See N. R. Bogatyrev and O. A. Bogatyrev, "TRIZ Evolution Trends in Biological and Technological Design Strategies," Proceedings of the 19th CIRP Design Conference—Competitive Design, March 30-31, 2009, https://dspace.lib.cranfield.ac.uk/bitstream /1826/3728/3/TRIZ_Evolution_Trends_in_Biological_and_Technological_Design _Strategies-2009.pdf.

119.  Vincent and Mann, "Systematic Technology Transfer," 172.

120.  This is more or less how Aristotle defines chance. See Aristotle, *Physics*, 44-46.

121.  Philip Ball, "Life's Lessons in Design," *Nature* 409 (2001): 413.

122.  Whether or not this or that biodesigner practices biomimicry depends on whether their use of biomimetic materials is modeled on nature, for an innovation only counts as bio-mimetic if it *intentionally* imitates nature. The present theory, it follows, does not simply integrate all biodesign into biomimicry in one fell swoop, but rather provides a theoreti-cal framework in which biodesigners could more readily embrace biomimicry, for they could come to see and understand their integration of living beings as modeled on nature.

## 3. NATURE AS MEASURE

1.  Immanuel Kant, *Groundwork of the Metaphysic of Morals*, trans. H. J. Paton (New York: Harper & Row, 1964).

2.  This basic idea is present already in *On the Genealogy of Morality*, but the exact expres-sion "the reevaluation of all values" was the intended title of a four-part work, only the first volume of which was completed.

3.  Jean-Paul Sartre, *L'existentialisme est un humanisme* (Paris: Gallimard, [1946] 1996).

4.  Max Horkheimer, *Eclipse of Reason* (London: Bloomsbury, [1947] 2004).

5.  Latour at one point explicitly affirms: "Everything is the measure of everything else" ("Toute chose est la mesure de toutes les autres"). Bruno Latour, "Irréductions," in *Pas-teur: guerre et paix des microbes* (Paris: Découverte, [1984] 2001), 243.

6.  This is why Latour calls his position "irreductionism."

7.  Janine Benyus, *Biomimicry: Innovation Inspired by Nature* (New York: Harper Perennial, 1997), epigraph, n.p.

8.  Benyus, *Biomimicry*, epigraph.

9.  By saying that nature contains knowledge, I am not affirming that this knowledge is consciously understood. An in-depth epistemological analysis of the claim that nature is capable of accumulating knowledge will be carried out in the next chapter.

10.  J. K. Feibleman, "Pure Science, Applied Science, and Technology: An Attempt at Defi-nitions," in *Philosophy and Technology: Readings in the Philosophical Problems of Technology*, 2nd ed., ed. Karl Mitcham and Robert Mackey (New York: Free Press, 1983), 37.

11.  I. C. Jarvie, "Technology and the Structure of Knowledge," in Mitchem and Mackey, eds., *Philosophy and Technology*, 55.

12.  Mario Bunge, "Toward a Philosophy of Technology," in Mitchem and Mackey, eds., *Philosophy and Technology*, 68.

13.  Henry Skolimowski, "The Structure of Thinking in Technology," in Mitchem and Mackey, eds., *Philosophy and Technology*, 46-49.

14.  To say that effectiveness (or efficiency) is a "technical" norm is not meant to imply that it is valid only within the sphere of technics. Effectiveness (of efficiency) may be a norm for action in general, as is the case when, walking from A to B, one takes the shortest route. But this is not to say that effectiveness is an ethical norm after all. Effective torture devices, for example, are not ethical on account of their effectiveness.

15.  Martin Heidegger, "Letter on Humanism," in *Basic Writings*, ed. D. F. Krell (Oxford: Routledge, 1993), 256.

16.  Heidegger, "Letter on Humanism," 258.

17.  Emmanuel Levinas, "Heidegger, Gagarin et nous," *Difficile liberté* (Paris: Albin Michel, 1963), 299-303.

18.  Levinas, "Heidegger, Gagarin et nous," 301.

19.  Levinas, 302.

20.  Harper Douglas, "Etymology of Ecology," Online Etymology Dictionary, https://www.etymonline.com/search?q=ecology (accessed July 12, 2022).

21.  Harper Douglas, "Etymology of Canon," Online Etymology Dictionary, https://www.etymonline.com/search?q=canon (accessed July 12, 2022).

22.  Benyus, *Biomimicry*, 7.

23.  Jackson himself observes that the principle was present—though not necessarily explicitly stated in as many words—first in the work of poets, such as Virgil and Shakespeare, and later on in the work of scientists, such as Albert Howard, a pioneer of organic farming (Wes Jackson, "Nature as Measure," in *Nature as Measure: The Selected Essays of Wes Jackson* (Berkeley: Counterpoint, [1994] 2011), 69-71). It would no doubt be possible to add various other important environmentalists, such as Thoreau or Leopold, to the "prehistory" of this concept.

24.  Jackson, "Nature as Measure," 78.

25.  The idea that the principle of nature as measure would, if it applied it all, apply to inter-human ethics explains why nature has yet to be taken as measure, or, rather, why the possibility of doing so has generally been so forcefully rejected. A classic statement of this view is provided by John Stuart Mill: "Either it is right that we should kill because nature kills; torture because nature tortures; ruin and devastate because nature does the like; or we ought not to consider at all what nature does, but what it is good to do." John Stuart Mill, "Nature," in *Three Essays on Religion* (London: Longmans, Green, Reader & Dyer, 1874), 31. "Conformity to nature," Mill concludes, "has no connection whatever with right and wrong" (62).

26.  As Jackson remarks: "We are not looking to nature for all the answers. The limits as well as the hideousness of social Darwinism of the last century, continuing into our own century, are well known." Jackson, "Nature as Measure," 78.

27.  Aristotle, *The Nicomachean Ethics*, trans. J. A. K. Thomson (London: Penguin, 2004), 149-51. See also Heidegger's commentary in *Plato's Sophist*, trans. R. Rojcewiz and A. Schuwer (Indianapolis: Indiana University Press, 2003), 34-36.

28. There may, for example, be certain relationships, such as relationships of care, in which the openness of human to fellow human plays an essential role, such that it would be unethical to replace the human carer with a robot, for the robot offers only a simulacrum of human openness: it reproduces the outward signs of openness to others, and of the care that such openness makes possible, but without actually being open to or caring for the person concerned.

29. In speaking of an "object" of ethics, I am using this term in its grammatical sense to designate that entity toward whom or which the ethical action of a "subject" is directed. It is certainly not meant to imply that the objects of ethical actions cannot also be subjects in their own right. My distinction between the subject and the object of ethical actions is thus akin to the more familiar distinction between moral agents and moral patients.

30. Bryan G. Norton, "Environmental Ethics and Weak Anthropocentrism," *Environmental Ethics* 6, no. 2 (1984): 131-48.

31. The thesis according to which sentience provides the ground for moral standing is usually called "sentientism." If, however, one understands the word "animal" in its original sense of an animated or sentient being, then "zoocentrism" and "sentientism" mean much the same thing. If I prefer to speak of "zoocentrism," it is simply because the word is more appropriate to the present discussion of object- or patient-centered ethics than is "sentientism."

32. The idea that sentience—or perhaps, rather, subjectivity—is the ultimate object of ethics has been defended by Tom Regan in *The Case for Animal Rights* (London: Routledge, 1983). And the idea that it is the *content* of sentience or subjectivity that is ethically important has been defended by Peter Singer in *Animal Liberation* (New York: Harper Collins, 1975).

33. See, for example, Paul Taylor, *Respect for Nature: A Theory of Environmental Ethics* (Princeton, NJ: Princeton University Press, 1986), 71-79.

34. Aldo Leopold, "The Land Ethic," in *A Sand County Almanac* (Oxford: Oxford University Press, 1949), 225.

35. Taylor, *Respect for Nature*, 206-12.

36. See, for example, Taylor, 200-206.

37. See Barbara Muraca, "The Map of Moral Significance," *Environmental Values* 20, no. 3 (2011): 375-96; and K. M. A. Chan, P. Balvanera, K. Benessaiah, M. Chapman, S. Díaz, E. Gómez-Baggethun, R. Gould et al., "Why Protect Nature? Rethinking Values and the Environment," *PNAS* 113, no. 6 (2016): 1462-65.

38. See Austin Himes and Barbara Muraca, "Relational Values: The Key to Pluralistic Valuation of Ecosystem Services," *Current Opinion in Environmental Sustainability* 35 (2018): 1-7.

39. Himes and Muraca, "Relational Values," 4.

40. See Bryan G. Norton, "Conservation and Preservation: A Conceptual Rehabilitation," *Environmental Ethics* 8, no. 3 (1986): 209-10.

41. See, for example, John Passmore, *Man's Responsibility for Nature: Ecological Problems and Western Traditions* (New York: Scribner's, 1974).

42. See Norton, "Conservation and Preservation," 200-1.

43. There is of course a danger that preserved nature, in the form of national parks and the like, comes to be valued solely for the entertaining days out it makes possible, and that

could just as easily be replaced by different days out, to, say, a theme park, cinema, or shopping center. But the fact that some people may relate to preserved nature in this primarily instrumental way does not alter the fact that other people—preservationists— see their relation to preserved nature as intrinsically valuable.

44.   It is important to note that the word "conservation" is often used today in a broader sense than in the late nineteenth century, such that it includes much of what previously fell under the distinct rubric of preservation. Conservation biology, for example, isn't only concerned with conserving nature as a resource for human beings.

45.   Kant, *Groundwork*, 96 (my emphasis).

46.   See, for example, Dwight A. Baldwin, Judith de Luce, and Carl Pletsch, eds., *Beyond Preservation: Restoring and Inventing Landscapes* (Minneapolis: University of Minnesota Press, 1994).

47.   See Baldwin, de Luce, and Pletsch, *Beyond Preservation*, 3-16.

48.   See Andrew Light and Eric S. Higgs, "The Politics of Ecological Restoration," *Environmental Ethics* 18, no. 3 (1996): 227-47.

49.   Andrew Light, "The Urban Blind Spot in Environmental Ethics," *Environmental Politics* 10, no. 1 (2001): 31.

50.   Robert Elliott calls this the "restoration thesis." See Robert Elliott, "Faking Nature," *Inquiry* 25, no. 1 (1982): 81.

51.   See John Basl, "Restitutive Restoration: New Motivations for Ecological Restoration," *Environmental Ethics* 32, no. 2 (2010): 135-47.

52.   See Eric Katz, "The Problem of Ecological Restoration," *Environmental Ethics* 18, no. 2 (1996): 222-24.

53.   The idea that there are four basic actions one may undertake with respect to nature, and also four corresponding "isms," is not entirely new. Frederick Turner has argued that the four "isms" are preservationism, conservationism, restorationism, and inventionism, with the latter according value above all to human creations and inventions, and therewith also to so-called novel ecosystems. Inventionism is not, however, an environmental ethic, in the sense of being an ethic governing our relation to nature, for its object is not nature at all, but, rather, human artifacts. So, whereas imitationism focusses on the imitation *of nature*, inventionism cannot focus on the invention *of nature*, for nature, by definition, cannot be invented by humans. On inventionism, see Frederick Turner, "The Invented Landscape," in Baldwin, de Luce, and Pletsch, eds., *Beyond Preservation*, 35-66.

54.   It is also interesting to consider here the French translations or equivalents of the word "biomimetics" (*la biomimétique*) and "biomimicry" (*le biomimétisme*). Whereas the suffix "-ics" (or -*ique* in French) in the word "biomimetics" (*biomimétique*) denotes a field of study (e.g., mathematics, hermeneutics, ballistics), the suffix "-ism" (-*isme* in French) denotes rather the existence of some sort of doctrine or set of values and beliefs.

55.   Benyus, *Biomimicry*, epigraph (my emphasis).

56.   Vincent Blok, "Earthing Technology: Towards an Ecocentric Concept of Biomimetic Technologies in the Anthropocene," *Techné: Research in Philosophy and Technology* 21 (2017): 127-49; Freya Mathews, "Biomimicry and the Problem of Praxis," *Environmental Values* 28, no. 5 (2019): 573-99.

57. Wes Jackson, *Consulting the Genius of Place: An Ecological Approach to a New Agriculture* (Berkeley: Counterpoint, 2010), 16.

58. Harper Douglas, "Etymology of Emulation," Online Etymology Dictionary, https://www.etymonline.com/word/emulation?ref=etymonline_crossreference (accessed July 12, 2022).

59. There are, of course, exceptions to this, most obviously ancient Greek Gaia worship.

60. It may at first seem that the idea that the good "presents itself" implies the existence of some sort of "ethical intuition" on the part of human subjects. In reality, however, there is no such implication. If we had some sort of in-built faculty of ethical intuition, then our understanding of the good would not change. But the goodness of Gaia is something that reveals itself historically. So, whereas in the former case the self-evident nature of the good derives from human beings, and more specifically from their faculty of ethical intuition, in the latter case the self-evident nature of the good derives rather from being or nature, which *reveals itself* as good.

61. I remarked in the introduction to this book that it is something of a commonplace to say that we are, and should see ourselves as, a part of nature. One meaningful way in which this claim is true is in the sense that we belong to and are part of Gaia, though it should not be forgotten that we also differ radically from Gaia's nonhuman inhabitants inasmuch as we alone inhabit the clearing.

62. See Emo Chiellini and Andrea Corti, "Oxo-Biodegradable Plastics: Who They Are and to What They Serve—Present Status and Future Perspectives," *Polyolefin Compounds and Materials*, ed. Mariam Al-Ali AlMa'adeed and Igor Krupa (New York: Springer, 2016), 341–54.

63. See Ren Wei and Wolfgang Zimmerman, "Microbial Enzymes for the Recycling of Recalcitrant Petroleum-Based Plastics: How Far Are We?," *Microbial Biotechnology* 10, no. 6 (2017): 1308–22.

64. Like Leopold ("Land Ethic," 225), I add the verb "tends" here because, while we cannot always know whether the consequences of this or that action will be beneficial, we can know that the consequences of certain types of action will be beneficial *as a rule of thumb*.

65. Taylor, for example, sets out various principles that may allow us to decide the outcomes of conflicts between humans and nature. See Taylor, *Respect for Nature*, 263–307.

66. Taylor argues that humans may permissibly destroy a portion of wild nature to meet their "non-basic" needs, such as building a museum, on the grounds that certain non-basic needs, such as cultural fulfillment, are "so great to them that they are unwilling to give them up" (287). But to justify killing huge swathes of living beings on the grounds that humans would be "unwilling" to forego a non-basic need that could be satisfied in this manner is hardly a rational argument deriving from the belief system of biocentrism. It is an attempt to align a theory of what is ethically right with what human beings are assumed to be willing to do.

67. An obvious objection to this position is that the human population is currently expanding and that, as new humans come into being, we have an ethical duty to provide them with the habitat they require, and this in turn necessitates the further development of human civilization into areas previously occupied by other organisms and species. But

while it is true that we have an ethical duty to provide any additional humans that come into being with the habitat they require, it does not follow that an increasing population justifies further intrusions into wild nature. Even where such basic needs as food are concerned, dietary changes, such as a transition to flexitarian, vegetarian or vegan diets, could easily enable even an expanded human population to inhabit the earth without destroying yet more wild nature.

68.  Richard Sylvan, "Is There a Need for a New Environmental Ethic?," in *Environmental Ethics: An Anthology*, ed. Andrew Light and Holmes Rolston (Oxford: Blackwell, 2003), 47–52.

69.  William Jordan, Robert L. Peters, and Edith B. Allen, for example, write: "By *restoration* we mean the recreation of entire *communities* of organisms, closely modeled on those occurring naturally." See "Ecological Restoration as a Strategy for Conserving Biological Diversity," *Environmental Management* 12, no. 1 (1988): 55.

70.  In the case of restoration, creative imitation could involve *composing* on the basis of models abstracted from the indigenous ecosystems, as well as *transforming* these models in the production of an artificial system that, in some respects, was deliberately different from the indigenous ones taken as models, for example, because it was also adapted to human recreational activity (e.g., by introducing footpaths or cycle paths).

71.  See, for example, Mark M. Brinson and Richard Rheinhardt, "The Role of Reference Wetlands in Functional Assessment and Mitigation," *Ecological Applications* 6 (1996): 69–76.

72.  See, for example, Taylor, *Respect for Nature*, 304–6.

73.  When I say that biomimetic ethics "centers" on imitation, I do not mean that it overlooks or even rejects the value of other environmental actions, but that it accords *primary significance* to imitation; it holds that it is *above all* by revolutionizing the products and systems we have made that we will be able to participate in the provision and maintenance of a habitable earth.

74.  J. Baird Callicott, *Thinking Like a Planet* (Oxford: Oxford University Press, 2013), 206–33.

75.  Callicott, *Thinking Like a Planet*, 236–37.

76.  Harper Douglas, "Etymology of Temper," Online Etymology Dictionary, https://www.etymonline.com/word/temper (accessed April 22, 2022).

77.  See Benyus, *Biomimicry*, 7.

78.  Jackson, *Consulting the Genius*, 16.

79.  Harper Douglas, "Etymology of Humility," Online Etymology Dictionary, https://www.etymonline.com/search?q=humility (accessed July 12, 2022); and Harper Douglas, "Etymology of Human," Online Etymology Dictionary, https://www.etymonline.com/word/human?ref=etymonline_crossreference#etymonline_v_16039 (accessed July 12, 2022).

80.  Harper Douglas, "Etymology of Arrogant," Online Etymology Dictionary, https://www.etymonline.com/word/arrogant (accessed April 22, 2022).

81.  Hölderlin, cited in Martin Heidegger, "The Question Concerning Technology," *Basic Writings*, ed. D. F. Krell (Oxford: Routledge, 1993), 333.

82.  Rachel Welchman, "The Virtues of Stewardship," *Environmental Ethics* 21, no. 4 (1999): 417.

83.  Norton, "Conservation and Preservation," 202.

## 4. NATURE AS MENTOR

1.  David Hume, *A Treatise of Human Nature* (London: Penguin, [1739-1740] 1985).

2.  Immanuel Kant, *Groundwork of the Metaphysic of Morals*, trans. H. J. Paton (New York: Harper & Row, [1785] 1964).

3.  G. E. Moore, *Principia Ethica* (Cambridge: Cambridge University Press, [1903] 1959).

4.  Friedrich Nietzsche, *On the Genealogy of Morality*, trans. M. Clark and A. J. Swensen (Indianapolis: Hackett, 1998), 12.

5.  This is not to say that Nietzsche rejects the idea of knowledge, only of universal knowledge that is independent of perspective. See Richard Schacht, "Nietzsche and the Perspectival," *Philosophical Topics* 33, no. 2 (2005): 193-225.

6.  Bruno Latour, "Irréductions," *Pasteur: guerre et paix des microbes* (Paris: Découverte, [1984] 2001), 323 (my translation).

7.  Bruno Latour, *Nous n'avons jamais été modernes* (Paris: Découverte, 1991).

8.  Bruno Latour, "When Things Strike Back—A Possible Contribution of 'Science Studies' to the Social Sciences," *British Journal of Sociology* 51, no. 1 (1999): 105-23.

9.  W. V. O. Quine, "Epistemology Naturalized," in *Knowledge: Readings in Contemporary Epistemology*, ed. Sven Bernecker and Fred Dretske (Oxford: Oxford University Press, 2000), 266-78.

10. Mark Rowlands, "Extended Cognition and the Mark of the Cognitive," *Philosophical Psychology* 22, no .1 (2009): 12-13.

11. Mark Rowlands, *Externalism: Putting Mind and World Back Together Again* (Chesham: Acumen, 2003), 173.

12. Clark, cited in Rowlands, *Externalism*, 19.

13. Aldo Leopold, "The Land Ethic," *A Sand County Almanac* (Oxford: Oxford University Press, 1949), 204.

14. It may perhaps be objected here that there are many instances of animals learning things from humans, including, for example, circus animals being taught how to perform tricks or sheepdogs learning how to herd sheep. But, while animals may be trained to behave in certain ways by humans, this is quite different from studying humans and seeking to learn from them, as humans do with respect to nature in biomimicry.

15. Donna Haraway, "Situated Knowledges: The Science Question in Feminism and the Privilege of Partial Perspective," *Feminist Studies* 14, no. 3 (1988): 575-99.

16. See *Gender/Body/Knowledge: Feminist Reconstructions of Being and Knowing*, ed. Alison M. Jaggar and Susan R. Bordo (New Brunswick, NJ: Rutgers University Press, 1989); and Elizabeth A. Grosz, *Volatile Bodies: Towards a Corporeal Feminism* (Bloomington: Indiana University Press, 1994).

17. See Karen Barbour, "Embodied Ways of Knowing: Revisiting Feminist Epistemology," in *The Palgrave Handbook of Feminism and Sport, Leisure, and Physical Education*, ed. L. Mansfield, J. Caudwell, B. Wheaton, and B. Watson (London: Palgrave Macmillan, 2018), 209-26.

18. See Margaret O. Little, "Seeing and Caring: The Role of Affect in Feminist Moral Epistemology," *Hypatia* 10, no. 3 (1995): 117-37.

19. Christopher Preston, "Epistemology and Environmental Philosophy: The Epistemic Significance of Place," *Ethics and the Environment* 10, no. 2 (2005): 1-4.

20. Christopher Preston, "Conversing with Nature in a Postmodern Epistemological Framework," *Environmental Ethics* 22, no. 3 (2000): 235.

21. Preston, "Epistemology and Environmental Philosophy," 238.

22. Benyus, *Biomimicry*, 7.

23. Christopher Preston, *Grounding Knowledge, Environmental Philosophy, Epistemology, and Place* (Athens: University of Georgia Press, 2003), 119.

24. Preston, *Grounding Knowledge*, 119–36.

25. Jim Cheney, "Postmodern Environmental Ethics: Ethics as Bioregional Narrative," *Environmental Ethics* 11, no. 2 (1989): 117–34.

26. Jim Cheney, "Universal Consideration: An Epistemological Map of the Terrain," *Environmental Ethics* 20, no. 3 (1998) 265.

27. Cheney, "Universal Consideration," 272.

28. Cheney, 272.

29. Cheney, 272.

30. Cheney, 273–74.

31. To be fair, Cheney certainly does see rocks as part of the earth (274). But if what we might call "epistemic considerability" is universal, then presumably we would also have to entertain the possibility of learning from rocks that are not part of the earth, such as asteroids.

32. Cheney, 268–70.

33. Robin Wall Kimmerer, *Braiding Sweetgrass: Indigenous Wisdom, Scientific Knowledge, and the Teachings of Plants* (London: Penguin, 2013).

34. This is not to say that wherever there is physis there is logos and epistēmē, for logos and epistēmē are absent from purely physical instances of physis, but rather that wherever logos is present, it is one with physis. It follows that when one takes purely physical instances of physis as model, one is not in fact learning from them, for they themselves possess no knowledge, but simply imitating various different traits that "ensist" within them. Creating energy by nuclear fusion is something stars do, but there is no knowledge of how to do so embedded in stars, for nuclear fusion did not arise through legein.

35. Humberto Maturana, "Introduction," in Humberto Maturana and Francisco Varela, *Autopoiesis and Cognition: The Realization of the Living* (Dordrecht: D. Reidel, 1980), xviii.

36. Immanuel Kant, *Critique of Pure Reason*, trans. Max Müller (London: Penguin, [1781, 1787] 2007), 124.

37. This is why Heidegger writes: "When we say that the lizard is lying on the rock, we ought to cross out the word 'rock' in order to indicate that whatever the lizard is lying on is certainly given in some way for the lizard, and yet is not known to the lizard as a rock." Martin Heidegger, *The Fundamental Concepts of Metaphysics: World, Finitude, Solitude*, trans. W. McNeill and N. Walker (Indianapolis: Indiana University Press, 1995), 198.

38. To give just one example, the overriding focus in social epistemology on testimony entails a certain marginalization of the study of the social transmission of knowing how.

39. It would also be possible to study the relation between biomimetic epistemology and ontology, in the sense of the study of physis. This field of study, which I have addressed only tangentially and sporadically, would include the question of how we come to have knowledge of physis, the limits of any knowledge we may have of physis, and how knowledge of physis differs from the more familiar case of traditional scientific knowledge about nature.

40.  Gilbert Ryle, "Knowing How and Knowing That," *Proceedings of the Aristotelian Society* 46 (1946): 1–16.

41.  See Jeremy Fantl, "Knowledge How," in *The Stanford Encyclopedia of Philosophy*, ed. E. N. Zalta, https://plato.stanford.edu/archives/fall2017/entries/knowledge-how/ (accessed June 29, 2022).

42.  There may be exceptions to this generalization. Dance, for example, is sometimes modeled on the rhythms and movements of nature, and the study of nature has also contributed to advances in mathematics.

43.  We may, of course, also learn how to do things from animals, as in the example offered by Benyus of learning what to eat and how to find it from primates (see Benyus, *Biomimicry*, 150–72). But this is a case of imitating the actions of another subject, not developing a biomimetic technology by abstracting and transferring over subjectless natural knowledge.

44.  The privilege accorded to "knowing that" is a legacy of logical positivism, which tends to focus on facts rather than things, on what is the case, rather than on what is.

45.  In the following discussion of biomimetic science, we will see that it is possible to divide the one overarching ethical measure, the way of being of Gaia, into various different submeasures, related to nutrient cycling, energy generation, and the like. But while one arrives at submeasures through delimitation and abstraction, these submeasures still differ from models in that they are not transferred over to technology (and so potentially also changed in the process), but remain that *against which* the ecological performance of our technologies is measured.

46.  Edmund Gettier, "Is Justified True Belief Knowledge?," in *Knowledge: Readings in Contemporary Epistemology*, ed. Sven Bernecker and Fred Dretske (Oxford: Oxford University Press, 2000), 13–15.

47.  J. Baird Callicott, John Van Buren, and Keith Wayne Browne, *Greek Natural Philosophy: The Presocratics and Their Importance for Environmental Philosophy* (San Diego, CA: Cognella, 2018), 239.

48.  See the section on Democritus's "Theory of Perception," in Sylvia Berryman, "Democritus," in *The Stanford Encyclopedia of Philosophy*, ed. E. N. Zalta, https://plato.stanford.edu/archives/win2016/entries/democritus/ (accessed June 29, 2022).

49.  Something like this idea had already been put forward by the Pythagoreans, but the model of being or nature they sought to express in mathematics was not *scientific* in nature, for, unlike the model of the atomists, it did not seek to represent the basic material components of reality and the interactions occurring between them.

50.  When I speak of models as being either confirmed or disconfirmed by measurements, I have in mind something less radical than either verification or falsification. Karl Popper famously argued that scientific hypotheses could never be verified, only ever falsified, for, while an accurate measurement that contradicted the model could prove it to be false, a measurement that was in accordance with the model could not prove it to be true, at least not universally. See Karl Popper, *The Logic of Scientific Discovery* (Abingdon-on-Thames: Routledge, [1959] 2002). But what this focus on falsification ignores is how measurements that conform to models significantly strengthen our belief in those models. When I say, then, that measurements may "confirm" the model, I do not mean that they may definitively verify it, but rather that they may significantly strengthen our

belief in it—in accordance with the etymology of the word "confirm" (from *con*-, meaning "with," and *firmare*, meaning "strengthen"). Conversely, when I say that a measurement "disconfirms" the model, I do not mean that it is definitively falsified by the measurement, but that our belief in the model is significantly weakened by the measurement.

51.    To highlight these two differences is not to deny that there may well be other differences between ancient and contemporary biomimicry. There was, for example, no drive among the ancients to use biomimicry as a source for ever more innovations. In Democritus and others, biomimicry is primarily an *explanation* for technology, an attempt to explain what it is and how it came into being, rather than a theory of *innovation*, of how new technologies might be developed.

52.    Harper Douglas, "Etymology of Ambiguous," Online Etymology Dictionary, https://www.etymonline.com/word/ambiguous (accessed April 25, 2022).

53.    For an example of this method, see Maibritt Pedersen Zari, "Ecosystem Services Analysis: Mimicking Ecosystem Services for Regenerative Urban Design," *International Journal of Sustainable Built Environment* 4, no. 1 (2015): 145-57.

54.    R. E. Blankenship, D. M. Tiede, J. Barber, G. W. Brudvig, G. Fleming, M. Ghirardi, M. R. Gunner et al., "Comparing Photosynthetic and Photovoltaic Efficiencies and Recognizing Potential for Improvement," *Science* 332 (2011): 805-9.

55.    See B. K. Meyer and P. J. Klar, "Sustainability and Renewable Energies—a Critical Look at Photovoltaics," *Physica Status Solidi* 5, no. 9 (2011): 318-23.

56.    Theocharis Tsoutsos, Niki Frantzeskaki, and Vassilis Gekas, "Environmental Impacts from the Solar Energy Technologies," *Energy Policy* 33 (2005): 289-96.

57.    Nicole C. McDonald and Joshua M. Pearce, "Producer Responsibility and Recycling Solar Photovoltaic Modules," *Energy Policy* 38 (2010): 7041-47.

58.    This may of course involve trade-offs. One may, for example, need to accept decreases in effectiveness in order to get closer to natural standards of appropriateness and sustainability.

59.    Laurence M. Peter, "Towards Sustainable Photovoltaics: The Search for New Materials," *Philosophical Transactions of the Royal Society A* 369 (2011): 1840-56.

60.    D. Larcher and J. M. Tarascon, "Towards Greener and More Sustainable Batteries for Electrical Energy Storage," *Nature Chemistry* 7 (2015): 19-29.

61.    It is interesting to note that German researchers have recently produced a hydrogen paste, made from combining hydrogen with various other elements, that resolves some of the problems associated with storing hydrogen in pure form. See "Powerpaste," Wikipedia, https://en.wikipedia.org/w/index.php?title=Powerpaste&oldid=1013576741 (accessed April 25, 2022).

62.    Thomas Kuhn, *The Structure of Scientific Revolutions*, 3rd ed. (Chicago: University of Chicago Press, 1996).

63.    J. Hwang, Y. Jeong, J. M. Park, K. H. Lee, J. W. Hong, and J. Choi, "Biomimetics: Forecasting the Future of Science, Engineering, and Medicine," *International Journal of Nanomedicine* 10 (2015): 5704.

64.    Airbus, for example, is designing aircraft based on models abstracted from bald eagles and long-eared owls, while also looking to snow geese as a model for more efficient

formation flying. "Biomimicry: A Fresh Approach to Aircraft Innovation," Airbus, https://www.airbus.com/en/newsroom/stories/2020-03-biomimicry-a-fresh-approach -to-aircraft-innovation (accessed April 25, 2022).

65. An important qualification would appear to be required here: that the natural phenomenon be precisely reproduced, for otherwise it is hard to see how it could be said to have been *fully* understood.

66. Andrew Pickering, "Asian Eels and Global Warming: A Posthumanist Perspective on Society and the Environment," *Ethics & Environment* 10, no. 2 (2005): 29–43.

67. This is particularly true when the concept of agency is extended to matter in general. Jane Bennett, for example, argues at length for the agency and even the "liveliness" of matter, but quite what difference it makes to either our technics or our ethics to see such items as a "black plastic work glove," a "white plastic bottle cap," and a "dead rat" as possessing an "energetic vitality" is not very clear. See Jane Bennett, *Vibrant Matter: A Political Ecology of Things* (Durham, NC: Duke University Press, 2010), 4–5.

68. Benyus, *Biomimicry*, 9.

69. Martin Heidegger, "Building, Dwelling, Thinking," in *Basic Writings*, ed. D. F. Krell (Oxford: Routledge, 1993), 363.

## CONCLUSION

1. Immanuel Kant, "What Is Enlightenment?," in *Kant: Political Writings*, ed. Hans Reiss, trans. H. B . Nisbett (Cambridge: Cambridge University Press, 1991), 54.

2. Kant, "What Is Enlightenment?," 54.

3. For an important recent contribution to the idea of nature as guide, see Ruth DeFries, *What Would Nature Do? A Guide for Our Uncertain Times* (New York: Columbia University Press, 2021).

4. See Henry Dicks, "A New Way of Valuing Nature: Articulating Biomimicry and Ecosystem Services," *Environmental Ethics* 39, no. 3 (2017): 286.

# Bibliography

Alam, Parvez. "Biomimetic Composite Materials Inspired by Wood." In *Wood Composites*, edited by Martin P. Ansell, 357–94. Cambridge: Woodhead, 2015.

Aldersey-Williams, Hugh. "Towards Biomimetic Architecture." *Nature Materials* 3 (2004): 277–79.

Ameisen, Jean-Claude. *La sculpture du vivant: le suicide cellulaire et la mort créatrice.* Paris: Seuil, 2003.

Aristotle. *The Nicomachean Ethics.* Translated by J. A. K. Thomson. London: Penguin, 2004.

—. *Physics.* Translated by Robin Waterfield. Oxford: Oxford University Press, 2000.

Bäck, Alan. *Aristotle's Theory of Abstraction.* Dordrecht: Springer, 2014.

Baldwin, D. A., J. de Luce, and C. Pletsch, eds. *Beyond Preservation: Restoring and Inventing Landscapes.* Minneapolis: University of Minnesota Press, 1994.

Ball, Philip. "Life's Lessons in Design." *Nature* 409 (2001): 413–16.

Barad, Karen. *Meeting the Universe Halfway: Quantum Physics and the Entanglement of Matter and Meaning.* Durham, NC: Duke University Press, 2007.

Barbour, Karen. "Embodied Ways of Knowing: Revisiting Feminist Epistemology." In *The Palgrave Handbook of Feminism and Sport, Leisure, and Physical Education*, edited by L. Mansfield, J. Caudwell, B. Wheaton, and B. Watson, 209–26. London: Palgrave Macmillan, 2018.

Basl, John. "Restitutive Restoration: New Motivations for Ecological Restoration." *Environmental Ethics* 32, no. 2 (2010): 135–47.

Bennett, Jane. *Vibrant Matter: A Political Ecology of Things.* Durham, NC: Duke University Press, 2010.

Bensaude-Vincent, Bernadette. "A Cultural Perspective on Biomimetics." In *Advances in Biomimetics*, edited by Anne George. InTech, 2011. https://www.intechopen.com/chapters/15671 (accessed July 20, 2022).

Bensaude-Vincent, B., H. Arribart, Y. Bouligand, and C. Sanchez. "Chemists and the School of Nature." *New Journal of Chemistry* 26 (2002): 1–5.

Benyus, Janine. *Biomimicry: Innovation Inspired by Nature.* New York: Harper Perennial, 1997.

Berryman, Sylvia. "Democritus." In *The Stanford Encyclopedia of Philosophy*, edited by E. N. Zalta. https://plato.stanford.edu/archives/win2016/entries/democritus/ (accessed June 21, 2022).

Bhushan, Bharat. "Biomimetics: Lessons from Nature—an Overview." *Philosophical Transactions of the Royal Society A* 367 (2009): 1445–86.

Black, Max. "Metaphor." *Proceedings of the Aristotelian Society* 55 (1954–1955): 273–94.

——. "More About Metaphor." *Dialectica* 31, nos. 3-4 (1977): 431–57.

Blankenship, R. E., D. M. Tiede, J. Barber, G. W. Brudvig, G. Fleming, M. Ghirardi, M. R. Gunner et al. "Comparing Photosynthetic and Photovoltaic Efficiencies and Recognizing Potential for Improvement." *Science* 332 (2011): 805-9.

Blok, Vincent. "Earthing Technology: Towards an Ecocentric Concept of Biomimetic Technologies in the Anthropocene." *Techné: Research in Philosophy and Technology* 21 (2017): 127-49.

Blok, Vincent, and Bart Gremmen. "Ecological Innovation: Biomimicry as a New Way of Thinking and Acting Ecologically." *Journal of Agricultural and Environmental Ethics* 29, no. 2 (2016): 203-17.

Blumenberg, Hans, and Anna Wertz. "Imitation of Nature: Toward a Prehistory of the Idea of the Creative Being." *Qui parle?* 12, no. 1 (2000): 28-29.

Bogatyrev, N. R., and O. A. Bogatyrev. "TRIZ Evolution Trends in Biological and Technological Design Strategies." Proceedings of the 19th CIRP Design Conference—Competitive Design, March 30-31, 2009. https://dspace.lib.cranfield.ac.uk/bitstream/1826/3728/3/TRIZ_Evolution_Trends_in_Biological_and_Technological_Design_Strategies-2009.pdf.

Braungart, Michael, and William McDonough. *Cradle to Cradle: Re-Making the Way We Make Things.* London: Vintage, 2009.

Brinson, Mark M., and Richard Rheinhardt. "The Role of Reference Wetlands in Functional Assessment and Mitigation." *Ecological Applications* 6 (1996): 69-76.

Brunetti, Jerry. *The Farm as Ecosystem: Tapping Nature's Reservoir—Geology, Biology, Diversity.* Greeley, CO: Acres USA, 2013.

Bryant, Levi, Nick Srnicek, and Graham Harman, eds. *The Speculative Turn: Continental Materialism and Realism.* Melbourne: re:Press, 2011.

Bunge, Mario. "Toward a Philosophy of Technology." In *Philosophy and Technology: Readings in the Philosophical Problems of Technology*, 2nd ed., edited by Karl Mitcham and Robert Mackey, 62-76. New York: Free Press, 1983.

Callicott, J. Baird. "A Neo-Presocratic Manifesto." *Environmental Humanities* 2, no. 1 (2013): 169-86.

——. *Thinking Like a Planet.* Oxford: Oxford University Press, 2013.

Callicott, J. B., J. Van Buren, and K. W. Browne. *Greek Natural Philosophy: The Presocratics and Their Importance for Environmental Philosophy.* San Diego, CA: Cognella, 2018.

Capra, Fritjof. *The Web of Life.* London: Harper Collins, 1997.

Cassirer, Ernst. "Form and Technology." In *The Warburg Years (1919–1933): Essays on Language, Art, Myth, and Technology*, translated by S. G. Loft and A. Calcagno, 272-316. New Haven, CT: Yale University Press, 2014.

Chan, K. M. A., P. Balvanera, K. Benessaiah, M. Chapman, S. Díaz, E. Gómez-Baggethun, R. Gould et al. "Why Protect Nature? Rethinking Values and the Environment." *PNAS* 113, no. 6 (2016): 1462-65.

Cheney, Jim. "Postmodern Environmental Ethics: Ethics as Bioregional Narrative." *Environmental Ethics* 11, no. 2 (1989): 117-34.

——. "Universal Consideration: An Epistemological Map of the Terrain." *Environmental Ethics* 20, no. 3 (1998): 265-77.

Chiellini, E., and A. Corti. "Oxo-biodegradable Plastics: Who They Are and to What They Serve—Present Status and Future Perspectives." In *Polyolefin Compounds and Materials*, edited by M. Al-Ali AlMa'adeed and I. Krupa, 341–54. New York: Springer, 2016.

Choay, Françoise. "La ville et le domaine bâti comme corps dans les textes des architectes-théoriciens de la première renaissance italienne." *Nouvelle revue de psychanalyse* 9 (1974): 239–51.

Christen, Yves. *L'animal est-il une personne?* Paris: Flammarion, 2011.

Codling, E. A., J. W. Pitchford, and S. D. Simpson. "Group Navigation and the 'Many Wrongs Principle' in Models of Animal Movement." *Ecology* 88, no. 7 (2007): 1864–70.

Commoner, Barry. *The Closing Circle: Nature, Man, and Technology.* New York: Knopf, 1971.

Coomaraswamy, Ananda. "The Nature of Medieval Art." *Studies in Comparative Religion* 15, nos. 1–2 (1983). http://www.studiesincomparativereligion.com/public/articles/The_Nature_of _Medieval_Art-by_Ananda_K_Coomaraswamy.aspx (accessed June 21, 2022).

Curd, Patricia. *The Legacy of Parmenides.* Las Vegas: Parmenides, 2004.

Cusa, Nicholas of. "Idiota da Mente." In *Nicholas of Cusa on Wisdom and Knowledge*, translated by Jasper Hopkins, 528–601. Minneapolis: Arthur J. Banning, 1996.

Dasgupta, D., ed. *Artificial Immune Systems and their Applications.* Berlin: Springer, 1999.

Dawkins, Richard. *The Selfish Gene.* Oxford: Oxford University Press, [1976] 1989.

Defries, Ruth. *What Would Nature Do? A Guide for Our Uncertain Times.* New York: Columbia University Press, 2021.

Dennett, Daniel. *Consciousness Explained.* Boston: Little, Brown, 1981.

Descola, Phillipe. *Par-delà nature et culture.* Paris: Gallimard, 2005.

Dessauer, Friedrich. *Philosophie der Technik. Das Problem der Realisierung.* Bonn: Cohen, 1927.

——. "Technology in its Proper Sphere." In *Philosophy and Technology: Readings in the Philosophical Problems of Technology*, 2nd ed., edited by Karl Mitcham and Robert Mackey, 317–34. New York: Free Press, 1983.

Dicks, H., J.-L. Bertrand-Krajewski, C. Ménézo, Y. Rahbé, J.-P. Pierron, and C. Harpet. "Applying Biomimicry to Cities: The Forest as Model for the City." In *Technology and the City: Towards a Philosophy of Urban Technologies*, edited by M. Nagenborg, T. Stone, M. G. Woge, and P. E. Vermass, 271–88. New York: Springer, 2020.

Dicks, Henry. "From Anthropomimetic to Biomimetic Cities: The Place of Humans in 'Cities Like Forests.'" *Architecture Philosophy* 3, no. 1 (2018): 65–68.

——. "A New Way of Valuing Nature: Articulating Biomimicry and Ecosystem Services." *Environmental Ethics* 39, no. 3 (2017): 281–99.

——. "The Philosophy of Biomimicry." *Philosophy & Technology* 29, no. 3 (2016): 223–43.

Dicks, Henry, and Vincent Blok, eds. "The Philosophy of Biomimicry and Bio-inspiration." Special issue, *Environmental Values* 28, no. 5 (2019).

Du, G., A. Mao, J. Yu, J. Hou, N. Zhao, J. Han, and Q. Zhao et al. "Nacre-Mimetic Composite with Intrinsic Self-Healing and Shape-Programming Capability." *Nature Communications* 10 (2019), doi: 10.1038/s41467-019-08643-x (accessed July 15, 2022).

Duns Scotus. *Ordinatio* 1.2.43. http://www.logicmuseum.com/wiki/Authors/Duns_Scotus /Ordinatio/Ordinatio_I/D2/Q2B (accessed July 20, 2022).

Dyson, Freeman. *Origins of Life.* 2nd ed. Cambridge: Cambridge University Press, 1999.

Elliott, Robert. "Faking Nature." *Inquiry* 25, no. 1 (1982): 81–93.

Eyth, Max. *Lebendige Kräfte: Sieben Vorträge aus dem Gebiete der Technik.* 4th ed. Berlin: J. Springer, 1924.

Fantl, Jeremy. "Knowledge How." In *The Stanford Encyclopedia of Philosophy,* edited by E. N. Zalta. https://plato.stanford.edu/archives/fall2017/entries/knowledge-how/ (accessed June 21, 2022).

Fayemi, P. E., K. Wanieck, C. Zollgrank, M. Maranzan, and A. Aoussat. "Biomimetics: Process, Tools, Practice." *Bioinspiration and Biomimetics* 12 (2017), doi: 10.1088/1748-3190/12/1/011002 (accessed July 15, 2022).

Feibleman, J. K. "Pure Science, Applied Science, and Technology: An Attempt at Definitions." In *Philosophy and Technology: Readings in the Philosophical Problems of Technology,* 2nd ed., edited by Karl Mitcham and Robert Mackey, 33–41. New York: Free Press, 1983.

Foltz, Bruce. *Inhabiting the Earth: Heidegger, Environmental Ethics, and the Metaphysics of Nature.* New York: Humanity, 1995.

Fox, Warwick. *Toward a Transpersonal Ecology: Developing New Foundations for Environmentalism.* Totnes: Resurgence, [1990] 1995.

Franssen, M., G. J. Lokhorst, and I. van de Poel. "Philosophy of Technology." In *The Stanford Encyclopedia of Philosophy,* edited by E. N. Zalta. http://plato.stanford.edu/archives/fall2015/entries/technology/ (accessed June 21, 2022).

Freeman, K. *Ancilla to the Pre-Socratic Philosophers (A Complete Translation of the Fragments in Diels, Fragmente der Vorsokratiker).* Cambridge, MA: Harvard University Press, 1948.

Fu, K., D. Moreno, M. Yang, and K. L. Wood. "Bio-Inspired Design: An Overview Investigating Open Questions from the Broader Field of Design-by-Analogy." *Journal of Mechanical Design* 136, no. 11 (2014), doi: 10.1115/1.4028289 (accessed July 15, 2022).

Gettier, Edmund. "Is Justified True Belief Knowledge?" In *Knowledge: Readings in Contemporary Epistemology,* edited by Sven Bernecker and Fred Dretske, 13–15. Oxford: Oxford University Press, 2000.

Girard, René. *Des choses cachées depuis la fondation du monde.* Paris: Grasset et Fasquelle, 1978.

Gould, S. J., and E. S. Vrba. "Exaptation—a Missing Term in the Science of Form." *Paleobiology,* 8, no. 1 (1982): 4–15.

Gleason, Henry A. "The Individualistic Concept of the Plant Association." *Bulletin of the Torrey Botanical Club* 53, no. 1 (1926): 7–26.

Grossman, Lisa. "Polystyrene Atoms Could Surpass the Real Deal." *New Scientist,* October 31, 2012. https://www.newscientist.com/article/dn22440-polystyrene-atoms-could-surpass-the-real-deal/.

Grosz, Elizabeth A. *Volatile Bodies: Towards a Corporeal Feminism.* Bloomington: Indiana University Press, 1994.

Gruber, P., D. Bruckner, H. B. Schmiedmayer, H. Stachelberger, and I. C. Gebeschuber. *Biomimetics—Materials, Structures, Processes.* Berlin: Springer, 2011.

Harman, Jay. *The Sharks Paintbrush: Biomimicry and How Nature Is Inspiring Innovation.* Ashland, OR: White Cloud Press, 2013.

Haraway, Donna. "Situated Knowledges: The Science Question in Feminism and the Privilege of Partial Perspective." *Feminist Studies* 14, no. 3 (1988): 575–99.

Hegel, G. W. F. *Hegel's Aesthetics, Lectures on Fine Art.* Vol.1. Translated by T. M. Knox. Oxford: Clarendon, 1975.

Heidegger, Martin. "Anaximander's Saying." In *Off the Beaten Track*, translated by J. Young and K. Haynes, 242–81. Cambridge: Cambridge University Press, [1950] 2002.

——. *Basic Problems of Phenomenology*. Translated by A. Hofstadter. Indianapolis: Indiana University Press, 1988.

——. *Basic Questions of Philosophy: Selected "Problems" of "Logic."* Translated by R. Rojcewicz and André Schuwer. Indianapolis: Indiana University Press, 1994.

——. *Basic Writings*. Edited by D. F. Krell. Oxford: Routledge, 1993.

——. *Being and Time*. Translated by John Macquarrie and Edward Robinson. Oxford: Blackwell, [1927] 1995.

——. *The Fundamental Concepts of Metaphysics: World, Finitude, Solitude*. Translated by W. McNeill and N. Walker. Indianapolis: Indiana University Press, 1995.

——. *Introduction to Metaphysics*. Translated by G. Fried and R. Polt. New Haven, CT: Yale University Press, [1953] 2000.

——. "On the Essence and Concept of *Physis* in Aristotle's *Physics B*, 1." In *Pathmarks*, 183–230. Translated by T. Sheehan. Cambridge: Cambridge University Press, 1998.

——. *Plato's Sophist*. Translated by R. Rojcewiz and A. Schuwer. Indianapolis: Indiana University Press, 2003.

——. *The Principle of Reason*. Translated by R. Lilly. Indianapolis: Indiana University Press, [1957] 1991.

Himes, Austin, and Barbara Muraca. "Relational Values: The Key to Pluralistic Valuation of Ecosystem Services." *Current Opinion in Environmental Sustainability* 35 (2018): 1–7.

Hobbes, Thomas. "Selections from the *Leviathan*." In *Philosophers Speak for Themselves: From Descartes to Locke*, edited by T. Smith and M. Grene, 157–229. Chicago: University of Chicago Press, 1967.

Hoffman, P. F., A. J. Kaufman, G. P. Halverson, and D. P. Schrag, "A Neoproterozoic Snowball Earth." *Science* 281 (1998): 1342–46.

Horkheimer, Max. *Eclipse of Reason*. London: Bloomsbury, [1947] 2004.

Howard, Albert. *An Agricultural Testament*. Mapusa: Other India, 1956.

Hume, David. *An Enquiry Concerning Human Understanding*. Oxford: Oxford University Press, [1748] 2007.

——. *A Treatise of Human Nature*. London: Penguin, [1739, 1740] 1985.

Hwang, J., Y. Jeong, J. M. Park, K. H. Lee, J. W. Hong, and J. Choi. "Biomimetics: Forecasting the Future of Science, Engineering, and Medicine." *International Journal of Nanomedicine* 10 (2015): 5701–13.

Iliadis, L. S., K. Kurkova, and B. Hammer. "Brain-Inspired Computing and Machine Learning." *Neural Computing & Applications* 32 (2020): 6641–43.

ISO 18458 Biomimetics—Terminology, Concepts, and Methodology. BSI (2015). https://www.iso .org/obp/ui/#iso:std:iso:18458:ed-1:v1:en (accessed June 21, 2022).

Jackson, Wes. *Consulting the Genius of Place: An Ecological Approach to a New Agriculture*. Berkeley: Counterpoint, 2010.

——. *Nature as Measure: The Selected Essays of Wes Jackson*. Berkeley: Counterpoint, [1994] 2011.

——. *New Roots for Agriculture*. Lincoln: University of Nebraska Press, 1980.

Jaggar, A. M., and S. R. Bordo, eds. *Gender/Body/Knowledge: Feminist Reconstructions of Being and Knowing*. New Brunswick, NJ: Rutgers University Press, 1989.

Jarvie, I. C. "Technology and the Structure of Knowledge." In *Philosophy and Technology: Readings in the Philosophical Problems of Technology*, 2nd ed., edited by K. Mitcham and R. Mackey, 54–61. New York: Free Press, 1983.

Jencks, Charles. *The Story of Postmodernism*. Chichester: John Wiley, 2011.

Jeong, K. H., J. Kim, and L. P. Lee. "Biologically Inspired Artificial Compound Eyes." *Science* 312, no. 5773 (2006): 557–61.

Jordan, W. R., III, R. L. Peters, and E. B. Allen. "Ecological Restoration as a Strategy for Conserving Biological Diversity." *Environmental Management* 12, no. 1 (1988): 55–72.

Jung, Y. H., P. Byeonhhak, J. U. Kim, and T. Kim. "Bioinspired Electronics for Artificial Sensory Systems." *Advanced Materials* 31 (2019), doi:10.1002/adma.201803637 (accessed July 15, 2022).

Kadochová, S., and J. Frouz. "Thermoregulation Strategies in Ants in Comparison to Other Social Insects, with a Focus on Red Wood Ants (*Formica rufa* Group)." *F1000Research* 2 (2014), doi: 10.12688/f1000research.2-280.v2 (accessed July 15, 2022).

Kant, Immanuel. *Critique of Judgment*. Translated by C. Meredith. Oxford: Oxford University Press, [1790] 2007.

—. *Critique of Pure Reason*. Translated by Max Müller. London: Penguin, [1781, 1787] 2007.

—. *Groundwork of the Metaphysic of Morals*. Translated by H. J. Paton. New York: Harper & Row, [1785] 1964.

—. "What Is Enlightenment?" Translated by H. B. Nisbett. In *Kant: Political Writings*, 2nd ed., edited by Hans Reiss, 54–60. Cambridge: Cambridge University Press, 1991.

Kaplinsky, Joe. "Biomimicry Versus Humanism." *Architectural Design* 76, no. 1 (2006): 66–71.

Kapp, Ernst. *Elements of a Philosophy of Technology: On the Evolutionary History of Culture*. Translated by L. K. Wolfe. Minneapolis: University of Minnesota Press, 2018.

Karthick, B., and R. Maheshwari. "Lotus-Inspired Nanotechnology Applications." *Resonance* 13, nos. 1141–45 (2008), doi: 10.1007/s12045-008-0113-y (accessed July 15, 2022).

Katz, Eric. "The Problem of Ecological Restoration." *Environmental Ethics* 18, no. 2 (1996): 222–24.

Kauffman, Stuart. *At Home in the Universe: The Search for the Laws of Self-Organization and Complexity*. Oxford: Oxford University Press, 1995.

Kimmerer, Robin Wall. *Braiding Sweetgrass: Indigenous Wisdom, Scientific Knowledge, and the Teachings of Plants*. London: Penguin, 2013.

Kostal, E., S. Stroj, S. Kasemann, M. Matylitsky, and M. Domke. "Fabrication of Biomimetic Fog-Collecting Superhydrophilic–Superhydrophobic Surface Micropatterns Using Femtosecond Lasers." *Langmuir* 34, no. 9 (2018): 2933–41.

Kuhn, T. *The Structure of Scientific Revolutions*. 3rd ed. Chicago: University of Chicago Press, 1996.

Kuttappan, S., D. Mathew, and M. B. Nair. "Biomimetic Composite Scaffolds Containing Bioceramics and Collagen/Gelatin for Bone Tissue Engineering—a Mini Review." *International Journal of Biological Macromolecules* 93, pt. B (2016): 1390–1401.

Larcher, D., and J. M. Tarascon. "Towards Greener and More Sustainable Batteries for Electrical Energy Storage." *Nature Chemistry* 7 (2015): 19–29.

Latour, Bruno. "Irréductions." In *Pasteur: guerre et paix des microbes*, 235–349. Paris: Découverte, [1984] 2001.

—. *Nous n'avons jamais été modernes*. Paris: Découverte, 1991.

—. *Politiques de la nature: comment faire entrer les sciences en démocratie*. Paris: Découverte, 1999.

—. "When Things Strike Back—a Possible Contribution of 'Science Studies' to the Social Sciences." *British Journal of Sociology* 51, no. 1 (1999): 105-23.

Lee, G. J., C. Choi, D. H. Kim, and Y. M. Song. "Bioinspired Artificial Eyes: Optic Components, Digital Cameras, and Visual Prostheses." *Advanced Functional Materials* 28, no. 24. (2017), doi: 10.1002/adfm.201705202 (accessed July 15, 2022).

Lefèvre, T., and M. Auger. "Spider Silk as a Blueprint for Greener Materials: A Review." *International Materials Reviews* 61, no. 2 (2016): 127-53.

Leopold, Aldo. "The Land Ethic." In *A Sand County Almanac*, 201-26. Oxford: Oxford University Press, 1949.

Levinas, Emmanuel. "Heidegger, Gagarin et nous." In *Difficile liberté*, 299-303. Paris: Albin Michel, 1963.

Light, Andrew. "The Urban Blind Spot in Environmental Ethics." *Environmental Politics* 10, no. 1 (2001): 7-35.

Light, Andrew, and Eric S. Higgs. "The Politics of Ecological Restoration." *Environmental Ethics* 18, no. 3 (1996): 227-47.

Little, Margaret O. "Seeing and Caring: The Role of Affect in Feminist Moral Epistemology." *Hypatia* 10, no. 3 (1995): 117-37.

Liu, Y., and G. Li. "A New Method for Producing 'Lotus Effect' on a Biomimetic Shark Skin." *Journal of Colloid and Interface Science* 388, no. 1 (2012): 235–42.

Lovelock, James. *Gaia: A New Look at Life on Earth*. Oxford: Oxford University Press, [1979] 2009.

Luhmann, Niklas. *Social Systems*. Stanford, CA: Stanford University Press, 1995.

Mann, Stephen. "Biomineralization and Biomimetic Materials Chemistry." *Journal of Materials Chemistry* 5 (1995): 935-46.

Mathews, Freya. "Biomimicry and the Problem of Praxis." *Environmental Values* 28, no. 5 (2019): 573-99.

—. "Can China Lead the World Towards an Ecological Civilization: A Manifesto." *Journal of Nanjing Forestry* 2 (2013). https://www.freyamathews.net/full-text-articles (accessed July 15, 2022).

—. "Towards a Deeper Philosophy of Biomimicry." *Organization & Environment* 24, no. 4 (2011): 364-87.

Maturana, Humberto, and Bernard Poulsen. *From Being to Doing: The Origins of the Biology of Cognition*. Heidelberg: Carl Auer, 2004.

Maturana, Humberto, and Francisco Varela. *Autopoiesis and Cognition: The Realization of the Living*. Dordrecht: D. Reidel, 1980.

McDonald, Nicole C., and Joshua M. Pearce. "Producer Responsibility and Recycling Solar Photovoltaic Modules." *Energy Policy* 38 (2010): 7041-47.

McKeag, Tom. "Auspicious Forms: Designing the Sanyo Shinkansen 500-Series Bullet Train." *Zygote Quarterly* 2 (2012): 14-35.

Meyer, B. K., and P. J. Klar. "Sustainability and Renewable Energies—a Critical Look at Photovoltaics." *Physica Status Solidi* 5, no. 9 (2011): 318-23.

Mill, John Stuart. "Nature." In *Three Essays on Religion*, 3-65. London: Longmans, Green, Reader, & Dyer, 1874.

Miller, Peter. *Smart Swarm*. London: Harper Collins, 2010.

Moore, G. E. *Principia Ethica*. Cambridge: Cambridge University Press, [1903] 1959.

Morin, Edgar. *La méthode, tome 1: la nature de la nature*. Paris: Seuil, 1977.

——. *La méthode, tome 2: la vie de la vie*. Paris: Seuil, 1980.

——. *Le paradigme perdu: la nature humaine*. Paris: Seuil, 1973.

Mourelatos, Alexander. *The Route of Parmenides*. Las Vegas: Parmenides, 2008.

Muraca, Barbara. "The Map of Moral Significance." *Environmental Values* 20, no. 3 (2011): 375–96.

Nietzsche, Friedrich. *On the Genealogy of Morality*. Translated by M. Clark and A. J. Swensen. Indianapolis, IN: Hackett, [1887] 1998.

Norton, Bryan G. "Conservation and Preservation: A Conceptual Rehabilitation." *Environmental Ethics* 8, no. 3 (1986): 190–220.

——. "Environmental Ethics and Weak Anthropocentrism." *Environmental Ethics* 6, no. 2 (1984): 131–48.

O'Rourke, J. M., and C. C. Seepersad. "Toward a Methodology for Systematically Generating Energy- and Materials-Efficient Concepts Using Biological Analogies." *Journal of Mechanical Design* 137, no. 9 (2015), doi:10.1115/1.4030877 (accessed July 15, 2022).

Owen, G. E. L. *Logic, Science, and Dialectic: Collected Papers in Greek Philosophy*. Ithaca, NY: Cornell University Press, 1986.

Oyen, Michelle. "Dreaming Big with Biomimetics: Could Future Buildings Be Made with Bone and Eggshells?" The Conversation, March 8, 2016, https://theconversation.com/dreaming-big-with-biomimetics-could-future-buildings-be-made-with-bone-and-eggshells-55739.

Packer, M. S., and D. R. Liu. "Methods for the Directed Evolution of Proteins." *Nature Review Genetics* 16 (2015): 379–94.

Passmore, John. *Man's Responsibility for Nature: Ecological Problems and Western Traditions*. New York: Scribner's, 1974.

Pedersen Zari, Maibritt. "Ecosystem Services Analysis: Mimicking Ecosystem Services for Regenerative Urban Design." *International Journal of Sustainable Built Environment* 4, no. 1 (2015): 145–57.

Peter, Laurence M. "Towards Sustainable Photovoltaics: The Search for New Materials." *Philosophical Transactions of the Royal Society A* 369 (2011): 1840–56.

Pickering, Andrew. "Asian Eels and Global Warming: A Posthumanist Perspective on Society and the Environment." *Ethics & Environment* 10, no. 2 (2005): 29–43.

Picon, Antoine. *Smart cities: théorie et critique d'un idéal auto-réalisateur*. Paris: Collection Actualités, 2013.

Plato, *The Republic*. Translated by Desmond Lee. London: Penguin, 1974.

Popper, Karl. *The Logic of Scientific Discovery*. Abingdon-on-Thames: Routledge, [1959] 2002.

Preston, Christopher. "Conversing with Nature in a Postmodern Epistemological Framework." *Environmental Ethics* 22, no. 3 (2000): 227–40.

——. "Epistemology and Environmental Philosophy: The Epistemic Significance of Place." *Ethics and the Environment* 10, no. 2 (2005): 1–4.

——. *Grounding Knowledge: Environmental Philosophy, Epistemology, and Place*. Athens: University of Georgia Press, 2003.

Quine, W. V. O. "Epistemology Naturalized." In *Knowledge: Readings in Contemporary Epistemology*, edited by S. Bernecker and F. Dretske, 266–78. Oxford: Oxford University Press, 2000.

Regan, Tom. *The Case for Animal Rights*. London: Routledge, 1983.

Ricoeur, Paul. *Temps et récit, tome 1: l'intrigue et le récit historique*. Paris: Seuil, 1983.

Rolston, Holmes, III. "Value in Nature and the Nature of Value." In *Philosophy and the Natural Environment*, edited by R. Attfield and A. Belsey, 13–30. Cambridge: Cambridge University Press, 1994.

Rowlands, Mark. "Extended Cognition and the Mark of the Cognitive." *Philosophical Psychology* 22, no. 1 (2009): 12–13.

—. "The Extended Mind." *Zygon* 44, no. 3 (2009): 628–41.

—. *Externalism: Putting Mind and World Back Together Again.* Chesham: Acumen, 2003.

Ryle, Gilbert. "Knowing How and Knowing That." *Proceedings of the Aristotelian Society* 46 (1946): 1–16.

Sandler, Ronald. *Character and Environment.* New York: Columbia University Press, 2007.

—. "A Theory of Environmental Virtue." *Environmental Ethics* 28, no. 3 (2006): 247–64.

Sartre, Jean-Paul. *L'existentialisme est un humanisme.* Paris: Gallimard, [1946] 1996.

Schacht, Richard. "Nietzsche and the Perspectival." *Philosophical Topics* 33, no. 2 (2005): 193–225.

Senanayake, R., and J. Jack. *Analogue Forestry: An Introduction.* Clayton: Department of Geography and Environmental Science, Monash University, 1998.

Shin, H., S. Jo, and A. G. Mikos. "Biomimetic Materials for Tissue Engineering." *Biomaterials* 24, no. 24 (2003): 4353–64.

Simondon, Gilbert. *Du mode d'existence des objets techniques.* Paris: Aubier, 1989.

Simons, Andrew M. "Many Wrongs: The Advantage of Group Navigation." *Trends in Ecology and Evolution* 19, no. 9 (2004): 453–55.

Singer, Peter. *Animal Liberation.* New York: HarperCollins, 1975.

Skolimowski, Henry. "The Structure of Thinking in Technology." In *Philosophy and Technology: Readings in the Philosophical Problems of Technology,* 2nd ed., edited by K. Mitcham and R. Mackey, 42–49. New York: Free Press, 1983.

Sylvan, Richard. "Is There a Need for a New Environmental Ethic?" In *Environmental Ethics: An Anthology,* edited by Andrew Light and Holmes Rolston, 47–52. Oxford: Blackwell, 2003.

Taylor, Paul. *Respect for Nature: A Theory of Environmental Ethics.* Princeton, NJ: Princeton University Press, 1986.

Todd, N. J., and J. Todd. *From Ecocities to Living Machines: Principles of Ecological Design.* Berkeley: North Atlantic, [1984] 1993.

Tomasello, Michael. *Origins of Human Communication.* Cambridge, MA: MIT Press, 2010.

Tsoutous, T., N. Frantzeskaki, and V. Gekas. "Environmental Impacts from the Solar Energy Technologies." *Energy Policy* 33 (2005): 289–96.

Turner, Frederick. "The Invented Landscape." In *Beyond Preservation: Restoring and Inventing Landscapes,* edited by D. A. Baldwin, J. de Luce, and C. Pletsch, 35–66. Minneapolis: University of Minnesota Press, 1994.

Varela, F., and P. Bourgine, eds. *Toward a Practice of Autonomous Systems: Proceedings of the First European Conference on Artificial Life.* Cambridge, MA: MIT Press, 1992.

Vincent, J. "Biomimetics—a Review." *Journal of Engineering in Medicine* 223, no. 8 (2009): 919–39.

Vincent, J., and D. Mann. "Systematic Technology Transfer from Biology to Engineering." *Philosophical Transactions of the Royal Society A* 360, no. 1791 (2002): 159–73.

Vincent, J., O. A. Bogatyrev, N. R. Bogatyrev, A. Bowyer, and A. K. Pahl. "Biomimetics: Its Practice and Theory." *Journal of the Royal Society Interface* 3, no. 9 (2006): 471–82.

Wahl, Daniel C. "Learning from Nature and Designing as Nature: Regenerative Cultures Create Conditions Conducive to Life." Biomimicry Institute, September 6, 2016. https:// biomimicry.org/learning-nature-designing-nature-regenerative-cultures-create-conditions -conducive-life/?mfsclkid=98d41688b4d411eca45905fd31d0a95e.

Ward, Frederick. *Discours prononcé à la séance d'ouverture du Congrès International de Bienfaisance.* Brussels: Librairie Européenne, C. Muquardt, 1856.

Wei, R., and W. Zimmerman. "Microbial Enzymes for the Recycling of Recalcitrant Petroleum-Based Plastics: How Far Are We?" *Microbial Biotechnology* 10, no. 6 (2017): 1308–22.

Welchman, Rachel. "The Virtues of Stewardship." *Environmental Ethics* 21, no. 4 (1999): 411–23.

Whitehead, A. N. *Process and Reality: An Essay in Cosmology.* New York: Free Press, 1978.

Wiener, Norbert, *Cybernetics: Or, Control and Communication in the Animal and the Machine.* New York: Technology, 1948.

——. *The Human Use of Human Beings: Cybernetics and Society.* Boston: Da Capo, [1950] 1954.

Wikisource. "Fragments of Parmenides." https://en.wikisource.org/w/index.php?title=Fragments _of_Parmenides&oldid=10779923 (accessed April 25, 2022).

——. "The *Poetics.* Translated by S. H. Butcher/1." https://en.wikisource.org/w/index.php?title=The _Poetics_translated_by_S._H._Butcher/1&oldid=12022729 (accessed April 22, 2022).

——. "Rhetoric (Freese)/Book 3." https://en.wikisource.org/w/index.php?title=Rhetoric_(Freese) /Book_3&oldid=3515996 (accessed April 25, 2022).

Yang, Y., X. Song, X. Li, Z. Chen, Q. Zhou, and Y. Chen. "Recent Progress in Biomimetic Additive Manufacturing Technology: From Materials to Functional Structures." *Advanced Materials* 30 (2018), doi: 10.1002/adma.201706539 (accessed July 15, 2022).

Young, K. J. "Mimicking Nature: A Review of Successional Agroforestry Systems as an Analogue to Natural Regeneration of Secondary Forest Stands." In *Integrating Landscapes: Agroforestry for Biodiversity Conservation and Food Sovereignty. Advances in Agroforestry,* vol. 12, edited by F. Montagnini, 179–209. Cham: Springer, 2017.

Zwart, Hub. "What Is Mimicked by Biomimicry? Synthetic Cells as Exemplifications of the Threefold Biomimicry Paradox." *Environmental Values* 28, no. 5 (2019): 527–49.

# Index

GPSR Authorized Representative: Easy Access System Europe, Mustamäe tee
50, 10621 Tallinn, Estonia, gpsr.requests@easproject.com

www.ingramcontent.com/pod-product-compliance
Lightning Source LLC
Chambersburg PA
CBHW022139020426
42334CB00015B/965